# Lecture Notes in Mathematics

Edited by A. Dold and B. Eckmann

T0253761

1309

Bernold Fiedler

# Global Bifurcation of Periodic Solutions with Symmetry

Springer-Verlag
Berlin Heidelberg New York London Paris Tokyo

**Author**

Bernold Fiedler
Institute of Applied Mathematics, University of Heidelberg
Im Neuenheimer Feld 294
D-6900 Heidelberg, Federal Republic of Germany

Mathematics Subject Classification (1980): 34XX, 35XX, 47H15, 58C27, 58F14, 58F22, 76E30, 80A30, 80A32

ISBN 3-540-19234-4 Springer-Verlag Berlin Heidelberg New York
ISBN 0-387-19234-4 Springer-Verlag New York Berlin Heidelberg

Library of Congress Cataloging-in-Publication Data. Fiedler, Bernold, 1956- Global bifurcation of periodic solutions with symmetry / Bernold Fiedler. p. cm.–(Lecture notes in mathematics; 1309) Bibliography: p. Includes index.
ISBN 0-387-19234-4 (U.S.)
1. Bifurcation theory. 2. Singularities (Mathematics) 3. Nonlinear operators. 4. Differential equations–Numerical solutions. 5. Differential equations, Partial–Numerical solutions. I. Title. II. Title: Periodic solutions with symmetry. III. Series: Lecture notes in mathematics (Springer-Verlag); 1309.
QA3.L28 no. 1309 [QA372] 510 s–dc 19 [514'.74] 88-12342


Printed in Germany

Printing and binding: Druckhaus Beltz, Hemsbach/Bergstr.
2146/3140-543210

für Marei

# Preface

The inherent harmony of periodic motions as well as of symmetry has exerted its own fascination, as it seems, ever since the dawn of thought. Today, such a "harmonia mundi" is at least hoped for on just about any possible scale: from elementary particle physics to astronomy.

In search of some harmony let us ask naive questions. Suppose we are given a dynamical system with some built-in symmetry. Should we expect periodic motions which somehow reflect this symmetry? And how would periodicity harmonize with symmetry?

These almost innocent questions are the entrance to a labyrinth of intricacies. Probing only along some fairly safe threads we are lead from dynamics to topology, algebra, singularity theory, numerical analysis, and to some applications. A global point of view will be one guiding theme along our way: we are mainly interested in periodic motions far from equilibrium.

For a method we rely on bifurcation theory, on transversality theory, and on generic approximations. As a reward we encounter known local singularities. As a central new aspect we study the global interaction and interdependence of these local singularities, designing a homotopy invariant. As a result, we obtain an index $\mathcal{H}$ which evaluates only information at stationary solutions. Nonzero $\mathcal{H}$ implies global Hopf bifurcation of periodic solutions with certain symmetries. Putting it emphatically, $\mathcal{H}$ harmonizes symmetry and periodicity . Curiously, $\mathcal{H}$ need not be homotopy invariant. It is one of my favorite speculations that this obstruction may hint at chaotic motions.

Cyclic motions relate to cyclic groups. Phrasing this relation between dynamics and algebra less sloppily: the symmetry of a periodic solution of a dynamical system is related to a cyclic factor within the group of symmetries of that system. Curiously, some period doubling bifurcations relate to the number 2, acting by multiplication on such a cyclic group. The multiplicative order of 2 relates to the number of possibly different indices $\mathcal{H}$ for a given system.

Symmetry, although beautiful, causes numerical difficulties. Basically, groups with irreducible representations of higher dimensions entail higher local singularities which are not very well understood. This is an obstacle to numerical pathfollowing algorithms. We will give a complete list of the easier, lower-dimensional generic bifurcations. Avoiding cyclic loops in the associated global bifurcation diagrams by a suitable homotopy invariant will be a central issue in our theoretical analysis. Both aspects are essential prerequisites for an efficient numerical pathfollowing method in dynamical systems with symmetries.

In real applications, as in real life, the lofty regions of harmony, periodicity, and symmetry are always confronted with the abysmal danger of destabilization. Surprisingly, there are still some applications where periodicity and symmetry is observed. We will concentrate on chemical waves as a model example below, though the theory is general. We obtain rotating waves (spirals) in continuous geometries, and phase-locked oscillations in discrete geometries.

Because it may not at all be easily detected by the reader, let me confess here a guiding principle for this book. Like so many others, I have tried to dismiss difficulty for beauty.

I happily say my thanks to everyone who has helped me. In particular, I would like to mention J. Alexander, G. Auchmuty, T. Bartsch, A. Brandis, S.-N. Chow, R. Cushman, R. Field, S. v. Gils, M. Golubitsky, W. Jäger, P. Kunkel, R. Lauterbach, J. Mallet-Paret, M. Marek, M. Medveď, J.-C. van der Meer, C. Pospiech, J. Sanders, D. Sattinger, R. Schaaf, A. Vanderbauwhede, A. Wagner, J. Yorke, and all those friends who have helped with proofreading. Typesetting the whole manuscript in TeX was a laborious task. It was performed by M. Torterolo with great patience. Finally, I am indebted to Springer-Verlag for an efficient and pleasant cooperation.

# Contents

# 1 Introduction

## 1.1 The question

We would like to find time-periodic solutions $x(t)$ of a nonlinear autonomous dynamical system

$$\dot{x}(t) = f(\lambda, x(t)), \qquad x \in X := \mathbb{R}^N, \quad f \in C^1 \quad . \tag{1.1}$$

In applications, such systems always contain parameters (coefficients). Let $\lambda \in \Lambda := \mathbb{R}$ denote one of them. Finding periodic solutions is usually more difficult than finding stationary, i.e. time-independent solutions $x(t) \equiv x_0$. Stationary solutions $(\lambda_0, x_0)$ satisfy

$$0 = f(\lambda_0, x_0). \tag{1.2}$$

Hopf bifurcation draws conclusions on periodic solutions of (1.1) from information on stationary solutions (1.2); and here and below we mean "nonstationary periodic" when we say periodic.

To describe local Hopf bifurcation suppose for a moment that $f(\lambda, 0) = 0$, for all real $\lambda$. Assume that the linearization $D_x f(\lambda, 0)$ at the stationary solution $(\lambda, 0)$ has a pair of simple eigenvalues

$$\lambda \pm i\beta(\lambda), \qquad \beta(\lambda) > 0 \tag{1.3}$$

for small $|\lambda|$. Then at least the linearized equation, at $\lambda = 0$,

$$\dot{y} = D_x f(0,0) y \tag{1.4}$$

has periodic solutions $y(t)$ of minimal period $2\pi/\beta(0)$. If $\pm i\beta(0)$ are the only purely imaginary eigenvalues of $D_x f(0,0)$, then the local Hopf bifurcation theorem, e.g. [Cra&Rab2], states that (1.1) with $f \in C^2$ has periodic solutions near $\lambda = 0$, $x = 0$. In fact, these periodic solutions form a continous branch and their minimal periods are close to $2\pi/\beta(0)$. Without a parameter $\lambda$, i.e. for fixed $\lambda = 0$, such a result could not hold in general.

The result above is called "local", because it only finds periodic solutions in some possibly very small neighborhood of $\lambda = 0$, $x = 0$. Global Hopf bifurcation finds periodic solutions which may be far away from the neighborhood where they originated. The first result in this direction is due to Alexander & Yorke [Ale&Y1], see §1.3 and in particular (1.29) for more details. Global Hopf bifurcation is our main concern here. Of course, global bifurcation implies local bifurcation.

Global as well as local bifurcation results require essentially some change of stability. Let us explain this with our previous example, $f(\lambda, 0) = 0$. Denote

$$E(\lambda) : \text{the number of eigenvalues of } D_x f(\lambda, 0) \text{ with strictly positive real part, counting algebraic multiplicity.} \tag{1.5}$$

In other words, $E(\lambda)$ is the unstable ("expanding") dimension of the stationary solution $(\lambda, 0)$. Then assumption (1.3) on the crossing of the pair of eigenvalues $\lambda \pm i\beta(\lambda)$ through the imaginary axis implies that $E(\lambda)$ changes by 2 as $\lambda$ increases through zero. We call this a "change of stability". Our principal goal will be an index $\mathcal{H}$ which evaluates changes of stability in such a way that $\mathcal{H} \neq 0$ implies global Hopf bifurcation.

We are interested in dynamical systems (1.1) with symmetry. Throughout we assume

$$\Gamma \text{ is a compact Lie group, acting orthogonally on } X := \mathbf{R}^N \text{ by a linear representation } \rho. \tag{1.6.a}$$

In other words:

$$\rho : \Gamma \to O(n)$$
$$\gamma \mapsto \rho(\gamma)$$

is a homorphism from the compact Lie group $\Gamma$ into the group $O(N)$ of orthogonal $N \times N$-matrices. See e.g. [Bre, Brö&tD, Sat&Wea] for generalities on Lie groups and representations. For practical purposes, we may assume that $\rho(\gamma) = id$ only for $\gamma = id$. This allows us to view $\Gamma$ as a closed subgroup of $O(N)$. A short-hand notation for the action of $\Gamma$ is $\gamma x := \rho(\gamma)x$, for $\gamma \in \Gamma$, $x \in X$. To tie up the group $\Gamma$ with our system (1.1), we require $f$ to be **equivariant** with respect to the action $\rho$ of $\Gamma$, i.e.

$$f(\lambda, \gamma x) = \gamma f(\lambda, x), \qquad \text{for all } \gamma \in \Gamma, \quad \lambda \in \mathbf{R}, \quad x \in \mathbf{R}^N. \tag{1.6.b}$$

Then (1.1) remains unchanged, if we replace $x$ by $\gamma x$. Thus, if $x(t)$ is a solution of (1.1), then $\gamma x(t)$ is also a solution, regardless which $\gamma \in \Gamma$ we choose. See e.g. [Sat1, Van1] for a reference on bifurcation theory for equivariant $f$.

If $x(t)$ is a periodic solution of system (1.1), then $\gamma x(t)$ may describe the same trajectory as $x(t)$ for suitably chosen $\gamma \in \Gamma$. In fact $\gamma$ could leave each point of $x(t)$ fixed, individually. Or $\gamma$ could leave the periodic orbit $\{x(t) \mid t \in \mathbf{R}\}$ fixed, as a set, possibly phase-shifting the individual points on it. In both cases we say that $\gamma$ belongs to the symmetry of the periodic solution $x(t)$. For more precision see §1.2, definition 1.1. This notion of symmetry leads us to our **principal question**:

How can we find periodic solutions with prescribed symmetry? (1.7)

For linear equivariant equations like (1.4), where $D_x f(0,0)$ has purely imaginary eigenvalues, we might find periodic solution and their symmetry explicitly, knowing the representation of $\Gamma$ on the eigenspace. For results on local Hopf bifurcation for nonlinear systems with symmetry see e.g. [Go&St1].

We approach question (1.7) from a global point of view. We design an index

$$\mathcal{H}^{\pm d}_{H_0, K_0} \tag{1.8}$$

such that nonzero $\mathcal{H}$ implies global Hopf bifurcation with certain possible symmetries. Again, $\mathcal{H}$ evaluates changes of stability of stationary solutions via purely imaginary eigenvalues in certain representation subspaces of X. For some more details see §1.4. A complete recipe is given in our main results: theorems 2.9 and 2.10 below.

Let us consider a first typical, but simple example: three identical, mutually coupled oscillators. Such examples go back to Turing [Tu]. With $x = (x_0, x_1, x_2)$, $x_j \in \mathbf{R}^{\tilde{n}}$, $x \in \mathbf{R}^{3\tilde{n}}$ our example may be written as

$$\begin{aligned} \dot{x}_0 &= \tilde{f}(x_0) + (x_2 - 2x_0 + x_1) \\ \dot{x}_1 &= \tilde{f}(x_1) + (x_0 - 2x_1 + x_2) \\ \dot{x}_2 &= \tilde{f}(x_2) + (x_1 - 2x_2 + x_0). \end{aligned} \tag{1.9}$$

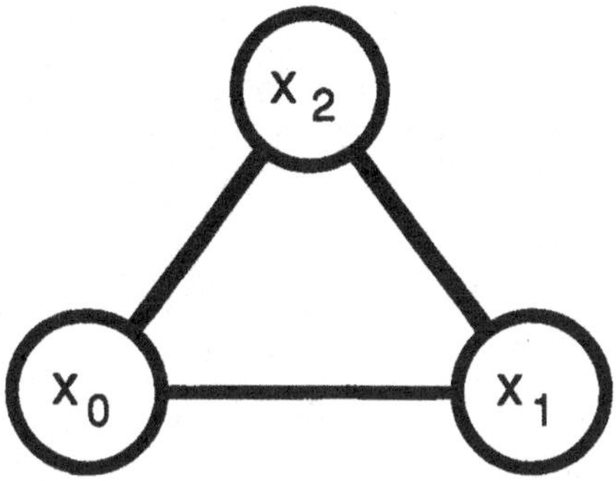


**Fig. 1.1** Three coupled oscillators

We suppress the parameter $\lambda$, here. In fig. 1.1 we depict system (1.9) as an equilateral triangle. The vertices stand for the oscillators $\dot{x}_0 =$ , $\dot{x}_1 =$ , $\dot{x}_2 =$ , and the sides represent "diffusive " coupling. System (1.9) remains invariant under any permutation of the indices $\{0, 1, 2\}$; the right hand side is equivariant under $\Gamma := S_3$, the symmetric group (permutations of three elements). From fig. 1.1 we see that $S_3$ is isomorphic to the dihedral group $D_3$, the group of orthogonal maps in the plane which leave an equilateral triangle invariant. System (1.9) could oscillate periodically in various ways: homogeneously ($x_0(t) \equiv x_1(t) \equiv x_2(t)$), with reflection symmetry ($x_0(t) \not\equiv x_1(t) \equiv x_2(t)$), with fixed phase-shifts over one third period between adjacent $x_j(t)$, with some other symmetry, or without any noticeable relation between the $x_j(t)$. Answering question (1.7), our index $\mathcal{N}$ will allow us a detailed global analysis of these phenomena, cf. §8.1. The first global results on such rings of coupled oscillators are due to Alexander & Auchmuty [Ale&Au2]. They rely on a topological result on global bifurcation of zeros of mappings with several (two) parameters [Ale1, Ale&Fitz].

Our approach to question (1.7) is more geometrically inclined. Motivated by the "snakes"-paper of Mallet-Paret & Yorke [M-P&Y1,2] we use generic, but equivariant approximations to the original problem (1.1). This will have the advantage that only a few types of bifurcations occur, and that global bifurcation diagrams can be understood systematically. We discuss this in §1.5 and, in excessive detail, in §§3,5-7,10. In [M-P&Y1,2], only the case of no symmetry, $\Gamma = \{id\}$, was considered. Another root of our approach is the elegant geometric treatment of local equivariant Hopf bifurcation by Golubitsky & Stewart [Go&Sch&St, Go&St1]. It inspired the very question (1.7), as well as our definition of symmetry of a periodic solution, and is behind the scene of most of our technical set-up.

Why should anyone be interested in a question like (1.7)? Our motivation is both "pure" and "applied". Symmetry prevails in many applied problems, e.g. oscillations in networks, in fluid dynamics, and in chemical reaction diffusion systems. A spectacular example are the rotating spirals in the Belousov-Zhabotinskii reaction, see fig. 1.2.

We devote §8 to such applications. Another "applied" goal is the development of quick, flexible tests which detect oscillations and give some indication of their spatio-temporal form in large distributed systems. Paradoxically, global results apply more easily than local results (but do not allow conclusions on stability, direction of bifurcation, etc.). As a "pure"

Fig. 1.2 A (clockwise) rotating spiral wave, courtesy of [Mü&Ple&Hess].

consequence we obtain local bifurcating branches for situations which could not be treated in [Go&St1], see theorem 9.1.

But local bifurcations, local singularities have been studied for quite a while now, even in equivariant settings. Our analysis adds a significant global feature: we investigate the interplay of these local singularities in global bifurcation diagrams. We believe that this global feature can and should be incorporated into other contexts as well. The problem of global Hopf bifurcation with symmetry just serves us as a model case.

Understanding the interplay of local singularities in global bifurcation diagrams usually uncovers some topological relations and restrictions, expressed by homotopy invariant indices. Knowing these global restrictions, as well as the basic local singularities, is in turn a prerequisite to the design of a successful numerical homotopy method for concrete applications. The simplest example is the monitoring of signs of determinants of the linearization, i.e. of Brouwer degree, to detect stationary bifurcation points; see e.g. [Deu&Fie&Kun]. This closes the circle of "pure" and "applied" motivations.

## 1.2 Symmetry of periodic solutions

Let us pin down what we mean by the symmetry of a periodic solution $x(t)$ of the $\Gamma$-equivariant differential equation (1.1). First we have to discuss "symmetries" of points $x \in X$. Given $x \in X$ the **isotropy** group $\Gamma_x$ of $x$ is defined as

$$\Gamma_x := \{\gamma \in \Gamma \mid \gamma x = x\}. \tag{1.10}$$

For example, consider the coupled oscillator system (1.9). If $x = (x_0, x_1, x_2)$ with $x_0 = x_1 = x_2$ then $\Gamma_x = \Gamma = S_3$. If $x_0 \neq x_1 = x_2$, then $\Gamma_x = \{id, (1\ 2)\} =: \langle(1\ 2)\rangle \cong \mathbb{Z}/2\mathbb{Z}$.

Conversely, given a subgroup $K$ of $\Gamma$ we may define the **fixed point subspace** $X^K$ of $X$ by

$$X^K := \{x \in X \mid Kx = x\}. \tag{1.11}$$

So $X^K$ consists of all elements $x$ of $X$ with isotropy $\Gamma_x$ at least $K$. In the example (1.9) we have $x \in X^\Gamma$ iff $x_0 = x_1 = x_2$, and $x \in X^{\langle(1\ 2)\rangle}$ iff $x_1 = x_2$.

This last example shows that $X^K$ may actually contain points $x$ with $\Gamma_x > K$. Throughout, we are interested in this typical case of a non-free group action, i.e. the conjugacy class of $\Gamma_x$ may depend on the choice of $x \neq 0$.

For solutions of (1.1) one would like to know $\Gamma_x$. The significant property of the linear subspaces $X^K$, on the other hand, is their flow invariance:

$$x(0) \in X^K \quad \text{implies} \quad x(t) \in X^K, \text{ for all } t. \tag{1.12}$$

Indeed, $x \in X^K$ implies $\dot{x} \in X^K$, because

$$K\dot{x} = Kf(\lambda, x) = f(\lambda, Kx) = f(\lambda, x) = \dot{x}.$$

Now consider a periodic solution $x(t)$ of (1.1) with minimal period $p > 0$. Let $C := \{x(t) \mid t \in \mathbb{R}\} \subset X$ denote the trajectory of $x(t)$. Then two relevant groups come to mind:

$$H := \{\gamma \in \Gamma \mid \gamma C = C\} \tag{1.13.a}$$

$$K := \Gamma_{x(t)} = \{\gamma \in \Gamma \mid \gamma x(t) = x(t)\}. \tag{1.13.b}$$

Note that $\Gamma_{x(t)}$ is in fact independent of $t$ because, by flow invariance of the spaces $X^K$, $x(0) \in X^{\Gamma_{x(t)}}$ and $x(t) \in X^{\Gamma_{x(0)}}$, i. e. $\Gamma_{x(0)} \geq \Gamma_{x(t)}$ and $\Gamma_{x(t)} \geq \Gamma_{x(0)}$. Thus $K$ is well-defined. Obviously, $K$ is a subgroup of the closed group $H$. For any $h \in H$, $x(t) \in C$, we have

$$h\, x(t) \quad = \quad x(t + \Theta(h)\, p). \tag{1.13.c}$$

Note that $\Theta(h) \in \mathbb{R}/\mathbb{Z}$ is defined independently of $t$. In fact $hx(t)$ solves the same differential equation (1.1) as $x(t)$ and the trajectories coincide as sets, by (1.13.a). Thus $hx(t)$ coincides with $x(t)$, up to a phase shift.

The obviously continuous map

$$\begin{aligned} \Theta: \quad & H \to \mathbb{R}/\mathbb{Z} \\ & h \mapsto \Theta(h) \end{aligned} \tag{1.14}$$

from $H$ to the (additive) group $\mathbb{R}/\mathbb{Z}$ is a homomorphism. Indeed

$$\begin{aligned} x(t + \Theta(h_1 h_2) \cdot p) &= h_1 h_2\, x(t) = h_1\, x(t + \Theta(h_2) \cdot p) = \\ &= x(t + (\Theta(h_1) + \Theta(h_2)) \cdot p), \qquad\qquad i.e. \\ \Theta(h_1\, h_2) &\equiv \Theta(h_1) + \Theta(h_2) \quad (mod\ \mathbb{Z}), \end{aligned}$$

because $p$ is the minimal period of $x(t)$. By definition, $ker\Theta = K$. By the homomorphism theorem [vdW, Lang], $K$ is a closed normal subgroup of the Lie group $H$ and

$$H/K \quad \cong \quad im\Theta \tag{1.15}$$

may be viewed as a closed subgroup of $\mathbb{R}/\mathbb{Z}$, cf. [Brö&tD, §1.4]. Let

$$\begin{aligned} \mathbb{Z}_n &:= \{0, \frac{1}{n}, \ldots, \frac{n-1}{n}\} \leq \mathbb{R}/\mathbb{Z}, \text{ for } n < \infty \\ \mathbb{Z}_\infty &:= \mathbb{R}/\mathbb{Z} \end{aligned} \tag{1.16.a}$$

denote the closed additive subgroups of $\mathbb{R}/\mathbb{Z}$. The cyclic groups $\mathbb{Z}_n$ should not be mixed up with the isomorphic groups

$$\begin{aligned} \mathbb{Z}(n) &:= \{0, 1, \ldots, n-1\} = \mathbb{Z}/n\mathbb{Z}, \\ \mathbb{Z}(\infty) &:= \mathbb{Z}. \end{aligned} \tag{1.16.b}$$

With this notation, (1.15) implies that

$$H/K \cong \begin{cases} \mathbb{Z}_n & \text{for some } n < \infty, \quad \text{or} \\ \mathbb{Z}_\infty \end{cases} \tag{1.17}$$

**1.1 Definition :**

*Let $x(t)$ be a periodic solution of system (1.1) with minimal period $p > 0$. We call the triple $(H, K, \Theta)$, defined by (1.13.a-c) above, the* **symmetry** *of $x(t)$.*

*Referring to (1.17) above, we call $x(t)$ a*

| | | |
|---|---|---|
| **concentric wave** | *if* | $H = K$ |
| **discrete wave** | *if* | $H/K \cong \mathbb{Z}_n\,,\quad 1 \leq n < \infty$ |
| **rotating wave** | *if* | $H/K \cong \mathbb{Z}_\infty\,.$ |

*Let $C \subseteq X$ be a set of stationary solutions of $f(\lambda, \cdot)$, i.e.*

$$f(\lambda, x) = 0 \qquad \text{for all } x \in C. \tag{1.18.a}$$

*We call $C$ a* **frozen wave,** *if there exists $x_0 \in C$ and subgroups $K := \Gamma_{x_0} \leq H \leq \Gamma$ such that the following two conditions hold:*

$$C = H \cdot x_0 \tag{1.18.b}$$

$$K = \Gamma_{x_0} \text{ is normal in } H \text{ and } H/K \cong \mathbb{Z}_\infty. \tag{1.18.c}$$

*We call the triple $(H, K, \pm\Theta)$ the* **symmetry** *of the frozen wave $C$, if $\Theta : H \to \mathbb{R}/\mathbb{Z}$ is any surjective homomorphism with kernel $K$.*

We comment on definition 1.1. First of all, it seems redundant to include $K = ker\Theta$ explicitly in the triple $(H, K, \Theta)$ which defines symmetry. Indeed, $H$ and $\Theta$ alone would suffice. Discussing secondary bifurcations it will be convenient though to nevertheless keep track of $K$ explicitly.

The terms concentric wave and rotating wave refer to the phenomenology of the Belousov-Zhabotinskii reaction in a petri dish, $\Gamma \geq SO(2)$. Concentric waves (alias target patterns), corresponding to $H = K = SO(2)$, are observed in form of circular rings propagating radially outwards in a time periodic fashion. For a snapshot of a rotating wave see fig. 1.2. Examples of discrete waves for the discrete symmetry of a triangular Turing ring, fig. 1.1, are discussed below. Note that concentric waves are somewhat degenerate examples of discrete waves.

In the symmetry $(H, K, \pm\Theta)$ of a frozen wave, the homomorphism $\Theta$ is determined only up to a sign. Indeed, $\Theta$ induces an isomorphism $H/K \to \mathbb{R}/\mathbb{Z}$, and the only continuous automorphisms of $\mathbb{R}/\mathbb{Z}$ are given by multiplication with $\pm 1$.

Condition (1.18.c) suggests that frozen waves are a pendant to rotating waves. Indeed, let $\mathcal{R}$ be the infinitesimal generator of the action of $H/K$ on $X^K$. In detail: we represent this action by orthogonal matrices, and obtain an isomorphism

$$\begin{aligned} \iota: \quad \mathbb{Z}_\infty &\to H/K \\ t &\mapsto exp(\mathcal{R}t). \end{aligned} \tag{1.19}$$

For some real $\alpha$, consider the transformation

$$y(t) \quad = \quad exp(-\alpha\mathcal{R}t)\ x(t) \tag{1.20}$$

on $X^K$. Then $y$ solves the equation

$$\dot{y}(t) = -\alpha\mathcal{R}\ y(t) + f(\lambda, y(t)) =: \hat{f}(\lambda, y(t)). \tag{1.21}$$

Choosing $\alpha = 1/p$, it turns out that $x(t)$ is a rotating wave for $f$ iff $H \cdot x(0)$ is a frozen wave for $\hat{f}$. The transformation (1.20) tells us that a rotating wave $x(t)$ "freezes", if viewed in a suitable rotating coordinate frame.

Conversely, let us start from a frozen wave $x \in C$ with symmetry $(H, K, \pm\Theta)$. Then the transformation (1.20) yields a rotating wave $y(t)$ with symmetry $(H, K, \Theta)$ or $(H, K, -\Theta)$, depending on the sign of $\alpha$. Viewing this as a perturbation result we may say that a rotating wave freezes and then starts rotating in the opposite direction, cf. definition 5.3 of a freezing, and theorem 5.11.

Viewed still differently, (1.20) and (1.21) tell us that $(\lambda, x(t))$ is a rotating wave if and only if $x_0 = x(0)$ with $\mathcal{R}x_0 \neq 0$ solves

$$0 = -\alpha\mathcal{R}x_0 + f(\lambda, x_0) \tag{1.21$'$}$$

for some $\alpha \neq 0$. On the other hand, $Hx_0$ is a frozen wave if and only if $x_0$ with $\mathcal{R}x_0 \neq 0$ solves $(1.21)'$ for $\alpha = 0$ .

Let us reinterpret symmetry of periodic solutions in an operator setting which is commonly used in global Hopf bifurcation. We rescale the minimal period $p$ of $x(t)$ to 1, defining

$$\xi(\tau) := x(p\tau) \quad . \tag{1.22}$$

Then $x(t)$ solves (1.1) iff $\xi$ solves

$$F(f, p, \lambda, \xi) := -\frac{1}{p}\dot{\xi} + f(\lambda, \xi) = 0 \quad . \tag{1.23}$$

Denoting the Banach spaces of continuous resp. once continuously differentiable functions with (not necessarily minimal) period 1 by $\tilde{C}^0$ resp. $\tilde{C}^1$, we may view $F(f, \cdot, \cdot, \cdot)$ for fixed $f$ as a map

$$F(f, \cdot, \cdot, \cdot): \quad \mathbb{R}^+ \times \mathbb{R} \times \tilde{C}^1 \ \to \ \tilde{C}^0 \ . \tag{1.24}$$

Fixing also $p, \lambda$, the map $F(f, p, \lambda, \cdot)$ is equivariant with respect to the action $\tilde{\rho}$ of $\tilde{\Gamma} := \Gamma \times S^1$ on $\tilde{\xi} \in \tilde{C}^0$ or $\tilde{C}^1$ defined by

$$(\tilde{\rho}(\gamma, \vartheta)\tilde{\xi})(\tau) \quad := \quad \rho(\gamma)\tilde{\xi}(\tau - \vartheta) \,, \tag{1.25}$$

where we write $S^1$ for the additive group $\mathbb{R}/\mathbb{Z}$.

We claim that $x$ has symmetry $(H, K, \Theta)$ iff $\xi$, defined by (1.22), has isotropy

$$\tilde{\Gamma}_\xi = H^\Theta := \{(h, \Theta(h)) \mid h \in H\}. \tag{1.26}$$

As before, it is understood that $K := ker\Theta$. To prove our claim, we follow the reasoning in [Go&St1, §6]. Applying the definition of symmetry of $x$, it is sufficient to show that $\tilde{\Gamma}_\xi = \tilde{H}^{\tilde{\Theta}}$ for some subgroup $\tilde{H}$ of $\Gamma$ and some homomorphism $\tilde{\Theta} : \tilde{H} \to S^1$. Let $\pi : \Gamma \times S^1 \to \Gamma$ denote projection onto the first factor and define $\tilde{H} := \pi(\tilde{\Gamma}_\xi)$. Then $\tilde{\Gamma}_\xi \cap ker\pi = \{id\}$, because $\xi$ has minimal period 1. Thus $\tilde{\Gamma}_\xi \cong \tilde{H}$, and we may hence write $\tilde{\Gamma}_\xi$ as $\tilde{H}^{\tilde{\Theta}}$ as was claimed above. Following [Go&St1, §6] we call $H^\Theta$ a twisted subgroup of $\Gamma \times S^1$ with twist $\Theta$.

Fixing an isomorphism from $\mathbb{Z}_n$ to $H/K$, we may represent the twist $\Theta$ by an integer $(mod\ n)$. Indeed, let $hK$ generate $H/K \cong \mathbb{Z}_n$ (assuming $n < \infty$) and fix $\iota$ to be given by

$$\begin{aligned} \iota : \quad & \mathbb{Z}_n \to H/K \\ & \tfrac{1}{n} \mapsto hK. \end{aligned} \tag{1.27}$$

Then

$$\Theta(h) \quad = \quad (\Theta \circ \iota)(\tfrac{1}{n}) \quad = \quad \Theta^* \,/\, n$$

for some $\Theta^* \in \mathbb{Z}(n)$. We will frequently identify $h$ with $1/n$ and $\Theta$ with $\Theta^* \in \mathbb{Z}(n)$, writing

$$\Theta(h) = \Theta \cdot h. \tag{1.28}$$

Using the isomorphism (1.19) instead of (1.27), the case $n = \infty$ is treated similarly. Representing $\Theta$ by integers is particularly convenient at secondary bifurcations of periodic solutions, where $\Theta$ may change, cf. §5.

We illustrate our symmetry terminology with the triangle of coupled oscillators (1.9), see fig. 1.1. Concentric waves, e.g., are periodic solutions $x(t)$ with $x_0(t) \equiv x_1(t) \equiv x_2(t)$. Their symmetry is $(H, K, \Theta) = (\Gamma, \Gamma, \Theta)$ where $\Gamma = S_3$. They satisfy $\dot{x}_0 = \tilde{f}(x_0)$, and diffusive coupling can be ignored altogether. Another example is given by $x_0(t) \not\equiv x_1(t) \equiv x_2(t)$ with $H = K = \langle(1\ 2)\rangle$, $\Theta = 0$. Such solutions satisfy

$$\begin{aligned} \dot{x}_0 &= \tilde{f}(x_0) + 2(x_1 - x_0) \\ \dot{x}_1 &= \tilde{f}(x_1) + (x_0 - x_1) \end{aligned}$$

and represent two asymmetrically coupled oscillators. A discrete wave could have symmetry $H = \langle(1\ 2)\rangle, K = \{id\}$, and $\Theta = 1$, which means $x_2(t) = x_1(t - \frac{p}{2})$ and $x_0(t) = x_0(t - \frac{p}{2})$. Such solutions are sometimes called standing waves. Another type of discrete waves satisfies $H = \langle(0\ 1\ 2)\rangle, K = \{id\}$, and $\Theta = 1$, which corresponds to

$$x_2(t) \equiv x_1(t - \frac{p}{3}) \equiv x_0(t - \frac{2p}{3}),$$

i.e. to fixed phase-differences between adjacent cells. Applying $(1\ 2) \in S_3$ to this solution we obtain a discrete wave with $\Theta = 2 \equiv -1\ (mod\ 3)$, i.e. rotation in the opposite direction. For specific examples of rotating and frozen waves see §8.2.

## 1.3 Some references

The literature on bifurcation problems is vast. We give some standard references to the field. Then we follow some of the threads to global bifurcation, concentrating on Hopf bifurcation. A more detailed attempt to put our results in perspective has to be postponed to §9. As a general reference to local bifurcation theory we mention the books by Chow & Hale [Chow&Ha], Golubitsky & Schaeffer [Go&Sch], Guckenheimer & Holmes [Gu&Ho], Iooss & Joseph [Io&Jo], as well as parts of Arnold [Arn3, ch.6], and Smoller [Smo, ch.13]. Bifurcations for iterates of maps are discussed e.g. in [Io]. Bifurcation theory for zeros of maps, viz. stationary solutions with several parameters, is known as singularity theory or catastrophe theory, see e.g. [Arn4, Arn&G-Z&Var, Go&Gui, Thom].

More specifically, local Hopf bifurcation is named after E. Hopf. In [Hopf], 1942, he proves the result which we have discussed in §1.1, assuming $x \in \mathbb{R}^N$ and $f$ analytic. His main motivation, though, was hydrodynamics. Hopf himself mentions Poincaré, who has considered the planar analytic Hamiltonian case being mainly motivated by periodically forced systems in celestial mechanics, cf. [Poi, ch.XXX], 1899. The general planar case was discussed extensively by Andronov and coworkers since 1929, see e.g. [And&Chai, And&Leo&Gor&Mai] and the note in [Arn3, p.271]. In 1977 a proof covering the infinite-dimensional case was given by Crandall & Rabinowitz [Cra&Rab2] in an analytic semigroup $C^2$-setting. They just relied on the implicit function theorem. Other modern accounts of local Hopf bifurcation, three of them based on center manifolds, are given e.g. in the books of Chow & Hale [Chow&Ha], Hassard & Kazarinoff & Wan [Has&Kaz&Wan], Iooss & Joseph [Io&Jo], and Marsden & McCracken [Mars&McCr].

The first global bifurcation result, concerning stationary solutions, is due to Rabinowitz [Rab]. Returning to the setting $f(\lambda, 0) = 0$ with unstable dimension $E(\lambda)$, as in (1.5), a version may be phrased as follows. If $E(\lambda)$ changes by an odd number, as $\lambda$ increases from $-\infty$ to $+\infty$, then an unbounded continuum of stationary solutions bifurcates from the trivial solution. The proof relies on degree theory, and we give a subjective version of it in §3; see also [Chow&Ha, §5.8] and [Smo, ch.13].

As we have mentioned above, the first result on global Hopf bifurcation without symmetry is due to Alexander & Yorke [Ale&Y1]; see also Ize [Ize1]. They both introduce period $p$ explicitly as a parameter. In the above setting, suppose $D_x f(0,0)$ is nondegenerate, and $D_x f(\lambda, 0)$ has some purely imaginary eigenvalues for $\lambda = 0$ but not for small $0 < |\lambda| \leq \varepsilon$. Assuming that

$$\frac{1}{2}(E(\varepsilon) - E(-\varepsilon)) \quad \text{is} \quad \text{odd}, \tag{1.29}$$

they obtain a global bifurcating continuum $\hat{C}$ of periodic solutions, by topological arguments involving stable homotopy theory. "Continuum" refers to the triple $(p, \lambda, \xi)$, and "global" means that $\hat{C}$ is unbounded or returns to some other bifurcation point on the trivial branch. Using Fuller index [Ful], Chow & Mallet-Paret & Yorke [Chow&M-P&Y1] later relaxed condition (1.29) to

$$\frac{1}{2}(E(\varepsilon) - E(-\varepsilon)) \quad \neq \quad 0. \tag{1.29$'$}$$

These results have one obvious and one subtle drawback. Obviously, we might not want to call $\hat{C}$ "global", if it remains bounded and just terminates at some other Hopf bifurcation point. It is a more subtle aspect to construct examples of continua in $(p, \lambda, \xi)$ which are unbounded, though $\lambda, \xi$ and minimal periods remain bounded. A concrete example for this important subtlety was constructed by Alligood & Mallet-Paret & Yorke [All&M-P&Y1],

cf. §3.4 and fig. 3.3 below. This is possible because $p$ in the operator setting (1.23) does not necessarily stand for minimal period. In fact, if $(p, \lambda, \xi)$ is a solution then $(kp, \lambda, \xi^k)$ is likewise a solution, if we define

$$\xi^k(\tau) := \xi(k\tau).$$

For a detailed discussion see §§3 and 9.3.

Both drawbacks have been circumvented at the expense of introducing the notion of "virtual periods", cf. definition 1.2 below and §4. For generic nonlinearities $f(\lambda, x)$ the drawbacks were fully remedied by Mallet-Paret & Yorke [M-P&Y1,2], who follow continua ("snakes") in $(\lambda, x)$ and simultaneously keep track of minimal period. Virtual periods, as introduced by Chow, Mallet-Paret, Yorke [M-P&Y2, Chow&M-P&Y2], arise if one approximates $f$ in (1.1) by generic nonlinearities. Following [Fie2], we give a detailed outline of this no-symmetry theory in §3 because it will be basic to our symmetry results.

Including symmetry, the books of Golubitsky, Schaeffer, Stewart [Go&Sch, Go&Sch&St], Sattinger [Sat1,2], and Vanderbauwhede [Van1,5] treat local bifurcations extensively. For a detailed study of local symmetry-breaking in elliptic equations see [Smo&Wa1-3, Van3, Van5]. Concerning local Hopf bifurcation with symmetry we have mentioned [Go&St1]. Rotating waves were also discussed, e.g., in [Au, Sche, Van2].

Global results are few in number. Globally-minded bifurcation of stationary solutions with symmetry was achieved by Cerami [Cer], Cicogna [Cic], and Pospiech [Pos1-3]. They all essentially pick a subgroup $K$ of $\Gamma$ and proceed along the global result of Rabinowitz [Rab] within the $f$-invariant subspace $X^K$. We could imitate this for periodic solutions, because $X^K$ is invariant under the flow (1.1). In $X^K$ the no-symmetry theorems from [Ale&Y1, Chow&M-P&Y1, Ize1, Ize2, Fie2] readily apply. For concentric waves ($H = K$, cf. §1.2) this approach is certainly appropriate. But it is not for $H > K$ : all information on $H$ and the action of $\Theta$ along the periodic solution will be lost completely. We are aware of only two previous results on global Hopf bifurcation with symmetry, which address this problem. Both are due to Alexander & Auchmuty: see [Ale&Au1] for rotating waves in a reaction diffusion system, and [Ale&Au2] for discrete waves in coupled oscillators.

However, these results are obtained via an operator setting similar to (1.23). They prescribe some symmetry $(\tilde{H}, \tilde{K}, \tilde{\Theta})$ for the periodic solutions $x(t)$, roughly as in definition 1.1, i.e.

$$h\, x(t) \quad = \quad x(t + \tilde{\Theta}(h)\tilde{p}), \quad \text{for all } h \in \tilde{H}, \tag{1.30}$$

but they do not know whether $\tilde{p}$ is the minimal period $p$ of $x(t)$, or just some multiple $kp$ of it. This way they obtain $H \geq \tilde{H}$, but no information on $\Theta$. In fact one can only conclude that

$$\Theta(h) \quad \equiv \quad k \cdot \tilde{\Theta}(h) \qquad (mod\ 1) \tag{1.31}$$

for some unknown $k$. For example, if $im\tilde{\Theta} \cong \tilde{H}/\tilde{K}$ is finite then $\Theta$ may be identically 0, picking $k = |\tilde{H}/\tilde{K}|$.

We are aiming at results which keep control of $\Theta$ and, at the same time, remedy the two drawbacks of the topology approach mentioned earlier. We will return to a comparison with the results of Alexander & Auchmuty in §9.4.

## 1.4 Virtual answers

Our notion of virtual symmetry is the key to our main results, summarized in theorems 2.9 and 2.10. "Mostly" virtual symmetry will coincide with symmetry, cf. definition 1.1. Virtual symmetry is defined as follows.

**1.2 Definition :**
*Let $x = x(t)$ be a stationary or a periodic solution of*

$$\dot{x}(t) = f(\lambda, x(t)) \quad . \tag{1.1}$$

*We call $q > 0$ a* **virtual period** *of $x$, and $(\hat{H}, \hat{K}, \hat{\Theta})$ a* **virtual symmetry** *of $x$, if there exists a solution $y$ of the linearized equation*

$$\dot{y}(t) = D_x f(\lambda, x(t))\ y(t) \tag{1.32}$$

*such that the pair $(x(t), y(t))$ has minimal period $q$ and symmetry $(\hat{H}, \hat{K}, \hat{\Theta})$ in the sense of definition 1.1; in particular*

$$\begin{aligned} \rho(h)\ x(t) &= x(t + \hat{\Theta}(h)\ q) \\ \rho(h)\ y(t) &= y(t + \hat{\Theta}(h)\ q) \end{aligned} \tag{1.33}$$

*for all $h \in \hat{H}$.*
*Similarly, suppose $f(\lambda, x) = 0$ and $y \in ker D_x f(\lambda, x)$ is such that the pair $(x, y)$ lies on a frozen wave $\hat{H} \cdot (x, y)$ with symmetry $(\hat{H}, \hat{K}, \pm\hat{\Theta})$ in the sense of definition 1.1, i.e. $\hat{K} = \Gamma_{(x,y)} := \Gamma_x \cap \Gamma_y$ is the isotropy of the pair $(x, y)$. Then we also call $(\hat{H}, \hat{K}, \pm\hat{\Theta})$ a virtual symmetry of $x$.*

The notion of virtual period is due to Chow, Mallet-Paret, and Yorke, see [M-P&Y2, Chow&M-P&Y2]. To be precise we should call $q$ a "virtual period of $x$ with respect to $f(\lambda, \cdot)$" etc., but for brevity we don't. Also, $\lambda$ is fixed in definition 1.2 and we might as well omit it.

Note that the minimal period $p > 0$ and the symmetry $(H, K, \Theta)$ of a periodic solution $x(t)$ are always a virtual period and a virtual symmetry of $x$, just putting $y \equiv 0$ or also $y \equiv \dot{x}$. Suppose $\dot{x}$ and its scalar multiples are the only periodic solutions of the variational equation (1.32). Then the minimal period is the only virtual period, and the symmetry is the only virtual symmetry of $x$. In particular this is the case for hyperbolic periodic solutions, i.e. for most "typical" periodic solutions. In general $x$ may have several, but finitely many, virtual periods and virtual symmetries. For stationary solutions $x_0$ the above remarks apply analogously. Note however, that a stationary solution $x_0$ has some virtual period and some virtual symmetry iff $D_x f(\lambda, x_0)$ has some purely imaginary nonzero eigenvalues, cf. lemma 4.8. Otherwise $(x, y)$ is necessarily stationary and its "minimal period" $q$ is not positive. For a thorough discussion of virtual symmetry see §4.

Next we describe at least the general flavor of our main results, theorems 2.9 and 2.10. For $\Gamma$-equivariant systems (1.1) we first fix any two closed subgroups $K_0 \leq H_0$ of $\Gamma$ such that $K_0$ is normal in $H_0$ and $H_0/K_0 \cong \mathbb{Z}_n$ is cyclic, $n \leq \infty$; the notation follows (1.16.$a$) above. A priori, these subgroups $H_0$, $K_0$ need not correspond to any symmetry $(H, K, \Theta)$ of any periodic solution at all. Next we pick a certain subset $d$ of $\mathbb{Z}(n)$, a so-called "binary orbit", cf. definition 2.4 and table 2.2. The set $d$ describes some maximal orbit in $\mathbb{Z}(n)$ under iterated multiplication by 2. Then we evaluate changes of unstable dimensions $E(\lambda)$ along stationary solution branches as $\lambda$ increases from $-\infty$ to $+\infty$, in reminiscence of conditions like (1.29), (1.29)'. These changes are counted in certain representation subspaces of $H_0/K_0 \cong \mathbb{Z}_n$ acting on $X^{K_0}$. The relevant representations are selected by the binary orbit $d$. This information is condensed into our integer-valued, global equivariant Hopf index

$$\chi^{\pm d}_{H_0,K_0} ,$$

cf. definition 2.8.

The principal conclusion is that

$$\chi^{\pm d}_{H_0,K_0} \neq 0 \quad \Rightarrow \quad \text{global Hopf bifurcation of periodic solutions with virtual symmetry at least } (H_0, K, \Theta),\ K_0 \leq K \leq H_0,\ \Theta \in d. \tag{1.34}$$

Here we work in the subspace $X^{K_0}$, where the cyclic group $H_0/K_0$ acts canonically, and the homomorphism $\Theta : H_0/K_0 \to S^1$ is represented by an integer $\Theta \in d$ as in (1.28).

By "global" we mean that there exists a continuum of periodic solutions in $\Lambda \times X^{K_0}$ which is unbounded, or which contains arbitrarily large virtual periods, cf. theorem 2.10. In (1.34), "at least" means that some virtual symmetry $(\hat{H}, \hat{K}, \hat{\Theta})$ satisfies

$$\hat{H} \geq H_0 \quad \text{and} \quad \hat{\Theta}|_{H_0} = 0. \tag{1.35}$$

Thus, by the isomorphism theorem [Lang, §I.4], $H_0/K$ is isomorphic to a subgroup of $\hat{H}/\hat{K}$ because

$$H_0/K \; = \; H_0/(H_0 \cap \hat{K}) \; \cong \; (H_0 \cdot \hat{K})/\hat{K} \; \leq \; \hat{H}/\hat{K}.$$

For rotating waves, i.e. for $H_0/K_0 \cong \mathbb{Z}_\infty$ and $\Theta \neq 0$, there even exists an unbounded continuum in $\Lambda \times X^{K_0}$. Periods do not enter, except that frozen waves are included in that continuum, cf. theorem 2.9. The theorems hold analogously for analytic semigroups, cf. corollary 2.13, and for integral equations, cf. the techniques in [Fie4]. For examples see §8, and §9 for further discussion. The discussion includes a proof that local Hopf bifurcation with maximal isotropy (in the sense of Golubitsky & Stewart [Go&St1]) always occurs, cf. theorem 9.1 in §9.1.

Commenting on our main result (1.34) we resume our discussion of virtual symmetry. Suppose that $x(t)$ is periodic and the minimal period $p$ is the only virtual period. We repeat, that this assumption is expected to hold for "most" periodic solutions. As a consequence, the symmetry $(\bar{H}, \bar{K}, \bar{\Theta})$ of $x$ extends any virtual symmetry $(\hat{H}, \hat{K}, \hat{\Theta})$ of $x$, i.e.

$$\bar{H} \geq \hat{H} , \qquad \bar{\Theta}|_{\hat{H}} = \hat{\Theta} ,$$

cf. lemma 4.11. For such solutions $x$ we may thus drop the word "virtual" in the statement (1.34) of our main result. We hasten to add that, in general, "virtual" cannot be dropped for virtual symmetries associated to non-minimal virtual periods.

Anyhow, there may exist suitable choices of $H_0$, $K_0$ and homomorphisms $\Theta \neq 0$ such that the triple $(H_0, K, \Theta)$ cannot be extended at all, for group theoretic reasons. We give examples in §8. Then at least "at least" can be dropped in (1.34). Note how important it is here to control $\Theta$ such that it remains nonzero, because $\Theta = 0$ extends trivially to any group $\hat{H} \geq H$.

## 1.5 Generic approximations

We obtain our results by generic approximation rather than by topological techniques. In this context genericity means the following. Fix closed subgroups $K_0 \leq H_0$ of the compact Lie group $\Gamma$ such that $K_0$ is normal in $H_0$ and $H_0/K_0 \cong \mathbb{Z}_n$ is cyclic, $n \leq \infty$, as before in §1.4. Note that $H_0$ leaves $X^{K_0}$ invariant. In fact, the normalizer $N(K_0)$ of $K_0$ leaves $X^{K_0}$ invariant, and $H_0$ is just a subgroup of $N(K_0)$. Still, we choose to ignore any symmetry above $H_0$, defining

$$\mathcal{F}: \quad \text{the space of } f \in C^4(\Lambda \times X^{K_0}, X^{K_0}) \text{ which are equivariant in the sense of (1.6.}b\text{) with respect to the representation of } H_0 \text{ on } X^{K_0}. \tag{1.36}$$

We endow $\mathcal{F}$ with the (weak) topology of uniform convergence on bounded sets. This makes $\mathcal{F}$ a Baire space [Di, Hirl], i.e. the countable intersection of open dense sets is still dense.

**1.3 Definition :**
*A subset $\mathcal{G}$ of $\mathcal{F}$ is called* **generic**, *if it contains a countable intersection of open dense subsets of $\mathcal{F}$. The elements of $\mathcal{G}$ are also called generic. A property is called generic, if it holds for a generic subset of $\mathcal{F}$.*

In §§5,10 we will prove that generically, i.e. for $f$ in a generic subset $\mathcal{G}$ of $\mathcal{F}$, only certain types of Hopf and secondary bifurcations can occur, cf. theorem 5.11. Just as the classical Kupka-Smale theorem [Kup, Smal], the proof relies on transversality theory or, more fundamentally, on Sard's theorem [Ab&Ro]. The complete classification is possible only because all local singularities, reflecting just $H_0/K_0 \cong \mathbb{Z}_n$ equivariance, are completely understood. This is the reason why we fix the groups $K_0, H_0$, a priori, possibly at the expense of ignoring symmetries above $H_0$.

Proving genericity results is somewhat involved, technically. But the advantages are obvious. We obtain a clear geometric picture of all possible global bifurcation diagrams. In particular, we can keep track of minimal periods and of symmetries (within $H_0/K_0$) at bifurcation points. We may even follow branches such that the minimal period jumps discontinuously, e.g. at period doubling bifurcations. This way we gain in flexibility over any purely topological approach.

As a second step we construct a homotopy invariant index $\Phi$ for periodic solutions, see definitions 6.1 and 6.4. This index tells us which branches we may select to find a global bifurcating continuum of periodic solutions, in the generic case, cf. theorems 2.6, 2.7, and §6. Also the index $\Phi$ carries information which describes the global interaction of the local singularities.

On the generic level, minimal periods and symmetries can still be controlled precisely. The notion of virtual period and virtual symmetry comes in if we want results for general, not necessarily generic, nonlinearities $f$. We have to approximate $f$ by generic $g_i$,

$$g_i \to f$$

and get corresponding approximating periodic solutions

$$x_i \to x \ .$$

It turns out that the limit of minimal periods and the limit of symmetries need not be a minimal period and a symmetry of the limiting $x$. Rather they are a virtual period and a virtual symmetry of $x$, cf. corollary 4.6.

For more details we refer the reader to §§3, 7. In §3 we give an extensive description of the genericity approach in the case of no symmetry ($\Gamma = \{id\}$). For a concrete, geometric study of the dynamic effects of a generic approximation on local Hopf bifurcation with $\Gamma = O(2), H_0 = SO(2), K_0 = \{id\}$ we recommend the interesting paper by v.Gils & Mallet-Paret [vG&M-P]. Of course, our generic approximations are more general but much less explicit.

## 1.6 A grasshoppers' guide

In §1 we give an introduction which is basic to all the rest. It should be read first, and completely.

The main results are stated in technically precise form in theorems 2.9 and 2.10. The basic definitions and assumptions are referenced there, to make these results directly accessible. For examples one may then skip §§3-7 and 10, looking at §8 instead and in particular at §8.3. It is also possible to directly jump into the discussion in §9. For open questions, see especially (9.7) and (9.13).

Another non-technical section is §3. We survey generic approximations in the case of no symmetry, indicating the analogous symmetry steps which occupy §§5-7 and 10. Our account of virtual symmetry, §4, can be read independently. But it is used freely in §§5-7 and 10, which constitute the technical center piece.

In §5, we develop the complete list of generic secondary bifurcations for $H_0/K_0 \cong \mathbb{Z}_n$ equivariant vector fields, cf. theorem 5.11 and table 5.2. Proofs are deferred mostly to §10 where some transversality machinery is applied to establish genericity. The generic global theory is then developed in §6. It includes the index $\Phi$ of periodic solutions, which describes the global interaction of local singularities. Taking theorem 5.11 for granted, §§6,7 can again be read pretty much independently of §§5,10. The approximation to nongeneric $f$ is carried out in §7. Virtual periods and virtual symmetries enter crucially, there.

For a fairly brief summary of some of the results we also refer to [Fie5]. We wish all grasshoppers an enjoyable "hop(f)".

# 2 Main results

## 2.1 Outline

In this section we summarize our main abstract results in a technically precise setting. For our general philosophy we refer to §1 and in particular to §1.4. We recall our distinction between generic nonlinearities $f$ and general "nongeneric" $f$, see §1.5, as well as the notions of concentric, discrete, rotating, and frozen waves with their symmetries $(H, K, \Theta)$, cf. definition 1.1. The four arising cases are treated according to table 2.1. For example, our main result on rotating/frozen waves for nongeneric $f$ is contained in theorem 2.9.

| | rotating and frozen waves | any waves |
|---|---|---|
| $f$ generic, §2.4 | theorem 2.6 | theorem 2.7 |
| $f$ nongeneric, §2.5 | theorem 2.9 | theorem 2.10 |

Table 2.1: Guide to main results

Before we can state these results, we need a definition of a global equivariant Hopf index $\mathcal{H}^d$ and, more basically, of the sets $d$ of integers which keep control of $\Theta$ along global branches. The sets $d$ are introduced in §2.3, definition 2.4. For $\mathcal{H}^d$ we give separate definitions for generic $f$ with only cyclic equivariance group $G = \mathbb{Z}_n$, $n \leq \infty$ ($\mathcal{H}_n^d$ : §2.4, definition 2.5) and for nongeneric $f$ with any compact equivariance group $\Gamma$ ($\mathcal{H}_{H_0,K_0}^{\pm d}$ : §2.5, definition 2.8). We conclude this section in §2.6 with some preliminary remarks on variants of our main results, postponing a more thorough discussion to §9.

Throughout §2 we fix two closed subgroups $K_0 \leq H_0 \leq \Gamma$ such that $K_0$ is a normal subgroup of $H_0$, and

$$H_0/K_0 \cong \mathbb{Z}_n = G, \quad n \leq \infty, \tag{2.1}$$

is a cyclic factor. We identify $G$ with $H_0/K_0$. We are interested in global Hopf bifurcation in $X^{K_0}$, and $G$ acts on $X^{K_0}$. To align the nongeneric case $(H_0, K_0)$ with the generic case $G$, notationally, it is convenient to consider

$$\dot{x} = f(\lambda, x) \tag{1.1}$$

restricted to the invariant subspace $x \in X^{K_0}$ and denote $X^{K_0}$ again by $X$. This way, we have the cyclic group $G = \mathbb{Z}_n$, $n \leq \infty$ acting on $X$ in both the generic and the nongeneric case. However, we explicitly write the global equivariant Hopf index as $\mathcal{H}_{H_0,K_0}^{\pm d}$ below, for the sake of clarity.

## 2.2 The generic center index

Throughout §§2.2-2.4 we fix the following setting. Let $G = \mathbb{Z}_n$, $n \leq \infty$, be a cyclic group with orthogonal representation $\rho$ on $X = \mathbb{R}^N$. For the nonlinearity $f$ in (1.1) we assume

$$f \in \mathcal{F}, \tag{2.2}$$

where the space $\mathcal{F}$ was defined in (1.36) and is understood with $H_0 = G$, $K_0 = \{id\}$. This means that $f \in C^4$ is $G$-equivariant with respect to $x \in X$. The highest regularity $C^4$ will not be used before §2.4, entering via §§5,10 into the main generic results. In §§2.2-2.3, $f \in C^3$ would be sufficient.

We begin to investigate generic local Hopf bifurcation by defining centers and generic centers. The term "generic" center will be justified later (theorem 5.2): for generic $f$ only generic centers occur.

**2.1 Definition :**

*Let $H$ be a closed subgroup of $G = \mathbb{Z}_n$. We call $(\lambda_0, x_0)$ an* **$H$-center** *if $f(\lambda_0, x_0) = 0$, $x_0 \in X^H$, and $D_x f(\lambda_0, x_0)$ has some purely imaginary nonzero eigenvalues. In other words: $(\lambda_0, x_0)$ is a stationary solution, invariant under $H$ (at least), with some virtual period (cf. definition 1.2).*

Suppose $(\lambda_0, x_0)$ is an $H$-center satisfying the following three additional conditions (2.3.$a$-$c$)

$D_x f(\lambda_0, x_0)$ has only one pair $\{\pm i\beta(\lambda_0)\}$ of purely imaginary eigenvalues and these eigenvalues are simple. (2.3.$a$)

In particular, 0 is not an eigenvalue and the stationary solutions near $(\lambda_0, x_0)$ form a local $C^4$-branch $(\lambda, x(\lambda))$ with continued eigenvalues $\alpha(\lambda) \pm i\beta(\lambda)$ of $D_x f(\lambda, x(\lambda))$; of course $\alpha, \beta$ are real, $\beta > 0$. The usual transverse crossing condition for local Hopf bifurcation then reads

$$D_\lambda \alpha(\lambda_0) \neq 0, \tag{2.3.b}$$

where $D_\lambda \alpha(\lambda_0)$ is the derivative of $\alpha(\lambda)$ with respect to $\lambda$ at $\lambda_0$. These two conditions ensure that local Hopf bifurcation occurs: there exists a local $C^3$-branch

$$s \mapsto (\lambda(s),\ x(s,\cdot),\ p(s)),\quad 0 \leq s < \varepsilon$$

of periodic solutions $x(s,\cdot)$ near $x_0$ with minimal period $p(s)$ near $2\pi/\beta(\lambda_0)$ and with $(x(s,t) - x_0)/s$ approaching, for $s \to 0$, an eigenfunction which corresponds to the eigenvalue $i\beta(\lambda_0)$ of $D_x f(\lambda_0, x_0)$; for a reference see e.g. [Cra&Rab2]. This branch contains **all** periodic solutions near $(\lambda_0, x_0)$. We finally require the curvature condition

$$D_s^2\, \lambda(0) \neq 0 \tag{2.3.c}$$

for the bifurcating branch. This condition ensures that periodic solutions are hyperbolic for $|s|$ small [Cra&Rab2].

**2.2 Definition :**

*We call $(\lambda_0, x_0)$ a* **generic $H$-center** *if it is an $H$-center satisfying the additional conditions (2.3.a-c). We call $(\lambda_0, x_0)$ a* **generic center** *if it is a generic $H$-center for some $H$.*

Given an $H$-center $(\lambda_0, x_0)$, the group $H$ acts on the joint generalized eigenspace of purely imaginary eigenvalues. Given a generic $H$-center, $H$ acts on the real two-dimensional eigenspace of $\{\pm i\beta(\lambda_0)\}$. We may identify this eigenspace with $\mathbb{C}$ in such a way that solutions $y$ of the linearized equation

$$\dot{y} = D_x f(\lambda_0, x_0)\, y$$

in this eigenspace are given by

$$y(t) = e^{i\beta(\lambda_0)t}\, y(0), \tag{2.4.a}$$

in complex notation. The action $\rho$ of $H$ on this eigenspace then takes the form

$$\rho(h)y = e^{2\pi i r h}\, y, \quad \text{for all } h \in H \leq \mathbb{Z}_n \leq \mathbb{R}/\mathbb{Z}, \tag{2.4.b}$$

where the integer $r$ (*mod* $n$) characterizes $\rho$. Note that our choice of $\beta(\lambda_0) > 0$ already determines the complexification of the eigenspace by (2.4.a) and hence $r$ is determined uniquely (*mod* $n$, if $n$ is finite) by (2.4.b) despite of the representations $r$ and $-r$ being (real) equivalent, cf. [Ser]. It turns out easily (cf. lemma 5.1 below) that the symmetry of the periodic solutions bifurcating from a generic $H$-center $(\lambda_0, x_0)$ is given by $(H, K, \Theta)$ with

$$\Theta \equiv r \qquad (mod\ |H|),$$

locally, provided that $H = G_{x_0}$ is the full isotropy of $x_0$. Here $\Theta$ is represented by an integer as was explained in (1.28).

This way the representations $r$ enter into our definition 2.5 of the global equivariant Hopf index $\mathcal{H}_n^d$. As a first step, we extract the relevant information on a change of stability, associated to this representation $r$, by a generic center index 中$^r$. The chinese symbol 中 for "center" or "China" is pronounced "zhong".

For simplicity, we specialize to the case that $(\lambda_0, x_0)$ is a generic $G$-center, i.e. a generic $H$-center with $H = G$. As before, let $(\lambda, x(\lambda))$ denote the corresponding local stationary branch with simple eigenvalues $\alpha(\lambda) \pm i\beta(\lambda)$, $\alpha(\lambda_0) = 0 < \beta(\lambda_0)$. The space $X^G$ of $G$-fixed vectors is invariant under the $G$-equivariant linearization $D_x f(\lambda_0, x_0)$. Denote

$E^0(\lambda_0)$ : the number of eigenvalues of $D_x f(\lambda_0, x_0)\,|_{X^G}$ with positive real part, counting algebraic multiplicity, (2.6)

$$\chi^{r'}(\lambda_0) := \begin{cases} \lim_{\varepsilon \searrow 0} \operatorname{sign}(\alpha(\lambda_0 + \varepsilon) - \alpha(\lambda_0 - \varepsilon)), & \text{for } r' = r, \\ 0, & \text{for } r' \neq r, \end{cases} \tag{2.7}$$

where $r$ denotes the representation associated to $\alpha(\lambda) \pm i\beta(\lambda)$ via (2.4) above. So the local crossing number $\chi^r(\lambda_0)$ is +1 (resp. −1) if the pair $\alpha(\lambda) \pm i\beta(\lambda)$ crosses the imaginary axis from left to right (resp. from right to left).

**2.3 Definition :**

*For a generic $G$-center $(\lambda_0, x_0)$ we define the* **center index** *by*

$$中^{r'}(\lambda_0, x_0) := (-1)^{E^0(\lambda_0)} \chi^{r'}(\lambda_0), \quad \text{for } 0 \leq r' < n. \tag{2.8}$$

*Note that* 中$^{r'} = 0$*, unless $r' = r$ is the representation of $G$ associated to $(\lambda_0, x_0)$.*

In case $G = \{0\}$, and hence $r = 0$, this center index reduces to the "no-symmetry" center index 中 which was introduced by Mallet-Paret & Yorke [M-P&Y1,2]. Recall that a brief account of the theory without symmetry will be given in §3.

## 2.3 Binary orbits

The set $d$ in $\mathcal{H}^{\pm d}_{H_0,K_0}$, $\mathcal{H}^d_n$ comes in through **2** ("period **doubling**"), acting by multiplication on $\mathbb{Z}(n) = \mathbb{Z}/n\mathbb{Z}$ :

$$\begin{aligned} \mathbf{2}: \quad \mathbb{Z}/n\mathbb{Z} &\to \mathbb{Z}/n\mathbb{Z} \\ r &\mapsto 2r \quad . \end{aligned} \tag{2.9}$$

This is motivated as follows. At a generic center we have $\Theta = r$ by lemma 5.1 below. But at secondary bifurcations the symmetries $(G, K, \Theta)$ and $(G, \tilde{K}, \tilde{\Theta})$ on the primary and secondary branch may differ from each other. In fact

$$\Theta \equiv \tilde{\Theta} \quad \text{or} \quad 2\tilde{\Theta} \qquad\qquad (mod\ n) \tag{2.10}$$

in the relevant cases; see §5 and especially corollary 5.13, (5.15) for more details. Seen from the centers, (2.10) describes some global interaction between generic centers with representations $r$, $\tilde{r}$ differing by a power of 2 $(mod\ n)$. We now give a formal definition of those subsets $d \subset \mathbb{Z}/n\mathbb{Z}$ for which a global equivariant Hopf index $\mathcal{H}^d_n$ will be defined below.

**2.4 Definition :**
*We define a relation* $\sim$ *on* $\mathbb{Z}(n) = \mathbb{Z}/n\mathbb{Z}$ *as follows:* $r_1 \sim r_2$ *iff there exist nonnegative integers* $j_1, j_2$ *such that*

$$2^{j_1} r_1 \quad \equiv \quad 2^{j_2} r_2 \qquad\qquad (mod\ n). \tag{2.11}$$

*The relation* $\sim$ *is an equivalence relation. Let* $D(n)$ *denote the set of equivalence classes. Equivalence classes* $d \in D(n)$ *are called* **binary orbits**.

So binary orbits $d$ are maximal (forward and backward) orbits under iterates of the multiplication map (2.9). Below, each choice of $d$ will give rise to a global Hopf index $\mathcal{H}^d_n$ and to a global result, separately, cf. our main results and our summary in §1.4, (1.34). Therefore we digress now to clarify the algebra of $D(n)$ and to compute the cardinality $|D(n)|$. We are indebted to A. Brandis for helpful discussions on this aspect.

We decompose $n$ into prime factors

$$n \;=\; 2^{\nu} n' \;=\; p_0^{\nu_0} \cdot p_1^{\nu_1} \cdot \ldots \cdot p_k^{\nu_k} \,,$$

where $n'$ is odd and the $p_j$ are distinct primes; $p_0 := 2$ and $\nu_o := \nu$. Correspondingly, we have a ring isomorphism

$$\iota: \qquad \mathbb{Z}(n) \quad \to \quad \mathbb{Z}(2^{\nu}) \times \mathbb{Z}(p_1^{\nu_1}) \times \cdots \times \mathbb{Z}(p_k^{\nu_k})$$

from $\mathbb{Z}(n)$ to the direct product of the $\mathbb{Z}(p_j^{\nu_j})$, see e.g. [Hasse, ch.4.2]. The projection onto $\mathbb{Z}(p_j^{\nu_j})$ is given by

$$r\ (mod\ n) \quad \mapsto \quad r\ (mod\ p_j^{\nu_j})\,.$$

Of course, the isomorphism $\iota$ induces a canonical bijection

$$D(n) \quad \to \quad D(2^{\nu}) \times D(p_1^{\nu_1}) \times \cdots \times D(p_k^{\nu_k})\,.$$

Note that $D(2^{\nu}) = \{\mathbb{Z}(2^{\nu})\}$ consists of a single element, and may thus be omitted. In particular

$$|D(n)| = |D(n')| = |D(p_1^{\nu_1})| \cdot \ldots \cdot |D(p_k^{\nu_k})|. \tag{2.12.a}$$

It remains to determine $|D(p_j^{\nu_j})|$ for odd primes $p_j$. Consider $\nu_j = 1$, first. Obviously

$$|D(p_j)| = 1 + \frac{p_j - 1}{ord^*_{p_j}(2)} ,$$

where $ord^*_{p_j}(2)$ denotes the order of 2 in the multiplicative group $\mathbb{Z}(p_j)^* = \{1, 2, \ldots, p_j - 1\}$. Any nonzero class $d \in D(n)$ has $ord^*_{p_j}(2)$ elements, in that case. The Artin conjecture claims that $|D(p_j)|$ attains its lower bound 2 for infinitely many primes; cf. e.g. [Sha, §§32 and 67].

Using [Hasse, ch.4.5] it is possible to prove that for odd prime numbers $p_j$

$$|D(p_j^{\nu_j})| \quad = \quad 1 + \sum_{l=1}^{\nu_j} \frac{p_j - 1}{ord^*_{p_j}(2)} \cdot gcd(p_j^{l-1}, \varsigma), \tag{2.12.b}$$

where $\varsigma$ is chosen such that

$$2^{p_j - 1} \equiv (p_j + 1)^{\varsigma}, \qquad mod\ p_j^{\nu_j},$$

and $gcd$ denotes the greatest common divisor. If the prime number $p_j$ is such that

$$2^{p_j - 1} \not\equiv 1, \qquad mod\ p_j^2,$$

then formula (2.12.$b$) simplifies to

$$|D(p_j^{\nu_j})| \quad = \quad 1 + \nu_j(p_j - 1)/ord^*_{p_j}(2) \quad = \quad 1 + \nu_j(|D(p_j)| - 1), \tag{2.12.b$'$}$$

because $\varsigma$ is prime to $p_j$, in that case. If $2^{p_j - 1} \equiv 1\,(mod\ p_j^2)$ then $p_j^2$ is called Wieferich square. Computer studies have shown that the only Wieferich squares with $p_j < 3 \cdot 10^9$ are given by $1093^2$ and $3511^2$, cf. [Sha, §§39 and 69]. For a relation of Wieferich squares to the restricted case of Fermat's Last Theorem see [Sha, §52]. In general, we may only estimate

$$\begin{aligned} |D(n)| &\geq \prod_{j=1}^{k} (1 + \nu_j \cdot (p_j - 1)/ord^*_{p_j}(2)) \\ &\geq \prod_{j=1}^{k} (1 + \nu_j). \end{aligned}$$

In particular $|D(n)|$ gets arbitrarily large. We omit proofs of (2.12.$b$), (2.12.$b$)$'$ for brevity. Instead, we refer to table 2.2. Some elements of $\mathbb{Z}(n)$ are represented by negative integers there.

For a concrete example, pick $n = 1986 = 2 \cdot 3 \cdot 331$. By table 2.2:

$$|D(1986)| = |D(3)| \cdot |D(331)| = 2 \cdot 12 = 24.$$

| $n$ | $\lvert D(n)\rvert$ | $d \in D(n)$ |
|---|---|---|
| 1 | 1 | (0) |
| 3 | 2 | (0) $(\pm 1)$ |
| 5 | 2 | (0) $(\pm 1, \pm 2)$ |
| 7 | 3 | (0) (1,2,4) $(-1,-2,-4)$ |
| 9 | 3 | (0) $(\pm 1, \pm 2, \pm 4)$ $(\pm 3)$ |
| 11 | 2 | (0) $(\pm 1, \pm 2, \pm 3, \pm 4, \pm 5)$ |
| 13 | 2 | (0) $(\pm 1, \pm 2, \pm 3, \pm 4, \pm 5, \pm 6)$ |
| 15 | 5 | (0) $(1,2,4,8)$ $(-1,-2,-4,-8)$ $(\pm 3, \pm 6)$ $(\pm 5)$ |
| 17 | 3 | (0) $(\pm 1, \pm 2, \pm 4, \pm 8)$ $(\pm 3, \pm 5, \pm 6, \pm 7)$ |
| 19 | 2 | (0) $(\pm 1, \pm 2, \pm 3, \pm 4, \pm 5, \pm 6, \pm 7, \pm 8, \pm 9)$ |
| 25 | 3 | (0) $(\pm 1, \pm 2, \pm 3, \pm 4, \pm 6, \pm 7, \pm 8, \pm 9, \pm 11, \pm 12)$ $(\pm 5, \pm 10)$ |
| 127 | 19 | (0) (1,2,4,8,16,32,64) etc. |
| 331 | 12 | (0) $(\pm 1, \pm 2, \pm 4, \pm 8, \pm 16, \pm 31, \pm 32, \pm 62, \pm 64, \pm 75, \pm 83,$ $\pm 124, \pm 128, \pm 150, \pm 166)$ etc. |
| $\infty$ | $\infty$ | (0) $(2^i\, r)_{i \geq 0}$, for any odd $r \in \mathbb{Z}$ |

Table 2.2: Some binary orbits

## 2.4 Generic global results

For a given $n$ we now define $|D(n)|$ possibly different global equivariant Hopf indices $\aleph_n^d$ ; one index for each $d \in D(n)$. We strengthen assumption (2.2), $f \in \mathcal{F}$, requiring genericity:

$$f \in \mathcal{F} \text{ is a generic, } G\text{-equivariant nonlinearity with only a finite number of } G\text{-centers.} \tag{2.13}$$

Here $G = \mathbb{Z}_n \leq \mathbb{R}/\mathbb{Z}$, $n \leq \infty$, as before. Assumption (2.13) looks somewhat mysterious since we do not specify any properties which $f$ is supposed to satisfy, at this stage. In §§5,10, however, we do specify such properties which are then proved to hold for generic nonlinearities $f \in \mathcal{F}$. Assumption (2.13) expresses these restrictions on $f$. For our notion of genericity see §1.5 and in particular definition 1.3.

**2.5 Definition :**

*Let genericity assumption (2.13) hold, and choose any binary orbit $d \in D(n)$. Then the* **generic global equivariant Hopf index** $\aleph_n^d$ *of the generic vector field $f$ is defined as*

$$\aleph_n^d := \sum_{r \in d} \Phi^r \quad , \tag{2.14}$$

*where the sum ranges over all $G$-centers (cf. definition 2.1) and the center index $\Phi^r$ is defined for generic $G$-centers (cf. definitions 2.2 and 2.3). For generic $f$, all $G$-centers are generic $G$-centers, cf. theorem 5.2 below. For binary orbits see definition 2.4.*

Using the terminology of definitions 1.1-1.3, 2.1-2.5 freely, we can now state our two main results on global $G$-equivariant generic Hopf bifurcation.

**2.6 Theorem :**

*Let the genericity assumption (2.13) hold for the cyclic group $G = \mathbb{Z}_\infty = \mathbb{R}/\mathbb{Z}$. Assume that*

$$\aleph_\infty^d + \aleph_\infty^{-d} \neq 0 \tag{2.15}$$

*for some nonzero binary orbit $d \in D(\infty) \setminus \{0\}$.*

*Then there exists an* **unbounded continuum** *$Z \subseteq \Lambda \times X$ consisting of generic $G$-centers, with representations $r \in d \cup (-d)$, and of rotating and frozen waves with symmetry*

$$(G, K, \Theta) \text{ and } \Theta \in d \cup (-d). \tag{2.16}$$

**2.7 Theorem :**

*Let the genericity assumption (2.13) hold for the cyclic group $G = \mathbb{Z}_n$, $n$ being finite or infinite. Assume that*

$$\aleph_n^d \neq 0 \tag{2.17}$$

*for some binary orbit $d \in D(n)$.*

*Then there exists a* **global continuum** *$Z \subseteq \Lambda \times X$ consisting of generic $G$-centers, with representations $r \in d$, and of periodic solutions with symmetry*

$$(G, K, \Theta) \quad \text{and} \quad \Theta \in d. \tag{2.16$'$}$$

*Here* **global** *means that*

*$Z$ is unbounded, or* (2.18.a)

*$Z$ contains periodic solutions with arbitrarily large minimal period.* (2.18.b)

The proof of these two results has to be postponed to §6 because it requires some insight into generic local bifurcations which will be accumulated in §§5,10. But an account of the case of no symmetry (i.e. $n = 1$), at least, will be given in §3 already.

At this stage, let us just compare the two generic results superficially. Theorem 2.7 is reminiscent of the "snakes" result by Mallet-Paret and Yorke [M-P&Y2, theorem 4.1] – except for the control on symmetry, of course. In particular minimal periods may become unbounded, with $Z$ remaining bounded in $\Lambda \times X$. For a discussion of this prominent aspect (2.18.$b$) of global Hopf bifurcation see §9.4 and the references given there. The assumption $G = \mathbb{Z}_\infty, d \neq \{0\}$ of theorem 2.6 singles out the case of rotating waves. In that case, theorem 2.7 of course still holds for either $d$ or $-d$ but periods might blow up. And then symmetry keeps control: the only way that period may blow up for a rotating wave is by decreasing its speed of rotation down to zero – it becomes a frozen wave, cf. §1.2. Theorem 2.7 will not carry us beyond the frozen wave: at this point the continuum $Z$ already becomes global because the (minimal) period blows up. But theorem 2.6 continues across the frozen wave. Changing $\lambda$ further, e.g., the frozen wave may start rotating again in the opposite direction. This way, centers with representations $r$ and $-r$ might interact. Therefore we consider the sum of the indices $\aleph_\infty^d$ and $\aleph_\infty^{-d}$ in assumption (2.15). Without anticipating the discussion in §9 we note here already, that rotating waves behave somewhat more like stationary solutions than like periodic solutions, globally, because period blow-up (2.18.$b$) can be circumvented.

## 2.5 Nongeneric global results

We now turn to the case of general, "nongeneric" $f \in C^1$, fixing new assumptions. Recall that $f$ is assumed to be equivariant with respect to the compact Lie group $\Gamma$, cf. (1.6), and $H_0/K_0 \cong \mathbb{Z}_n$, $n \leq \infty$, is a cyclic factor of $\Gamma$, cf. (2.1). We are aiming at periodic solutions with (virtual) symmetry at least $(H_0, K, \Theta)$, $K_0 \leq K \leq H_0$. Such solutions can bifurcate only from $H_0$-centers $(\lambda_0, x_0)$, i.e. from centers $x_0 \in X^{H_0}$, cf. lemma 4.8. We need some assumptions on the restricted linearizations

$$L^{K_0} \quad := \quad D_x f(\lambda, x)|_{X^{K_0}} \quad \in \quad \mathcal{L}(X^{K_0}, X^{K_0}) \tag{2.19}$$

at stationary solutions $(\lambda, x) \in \Lambda \times X^{K_0}$. We require

the set of stationary solutions in $\Lambda \times X^{K_0}$ consists of a finite number of branches (2.20.a)

$$(\lambda, x^l(\lambda)) \in \Lambda \times X^{H_0}, \quad l = 0, \cdots, \bar{l}$$

which are globally parametrized over $\lambda \in \Lambda$, and the eigenvalues of the linearization $L_l^{K_0}(\lambda)$ at $(\lambda, x^l(\lambda))$ are always nonzero;

the set of $H_0$-centers (with respect to $L^{K_0}$) is bounded; (2.20.b)

the linearization $L_l^{K_0}(\lambda)$ depends analytically on $\lambda$ near centers. (2.20.c)

The restrictive aspect of nondegeneracy assumption (2.20.a) is discussed in §9; cf. also remark 2.11. But assumptions (2.20.b, c) are not particularly aggravating in applications, cf. §8. Our analyticity assumption (2.20.c) will be motivated in §4 (cf. lemma 4.11).

Defining the global equivariant Hopf index $\mathcal{H}^{\pm d}_{H_0,K_0}$ for nongeneric $f$ requires some preparation. Recall that $H_0/K_0 \cong \mathbb{Z}_n = G$ acts canonically on $X^{K_0}$, and the restricted linearization $L^{K_0}$ commutes with this action. Identifying real two-dimensional spaces with $\mathbb{C}$, the real inequivalent irreducible types $\rho_r$ of real representations of $\mathbb{Z}_n$ are given in complex notation by

$$\begin{aligned} \rho_r(h)z \quad &= \quad e^{2\pi i r h} z, \qquad h \in \mathbb{Z}_n,\ z \in \mathbb{C}, \\ & \qquad\qquad\qquad\quad r \in \mathbb{Z},\ 0 \leq r \leq n/2. \end{aligned} \tag{2.21}$$

Of course, the cases $r = 0$, $r = n/2$ (for $n$ even) should be read for scalar real $z$, to be irreducible. For a background on these easy representations see again [Ser]. Mapping $z$ into $\bar{z}$, we note that $\rho_{-r}$ is real equivalent, but not complex equivalent, to $\rho_r$. These representations decompose $X^{K_0}$ uniquely into real representation spaces $X_r^{K_0}$

$$X^{K_0} \quad = \quad \bigoplus_{0 \leq r \leq n/2} X_r^{K_0}, \tag{2.22}$$

the representation on $X_r^{K_0}$ being given by a couple of copies of $\rho_r$. Again $L_l^{K_0}(\lambda)$ restricts to each $X_r^{K_0}$

$$L_l^r(\lambda) \quad := \quad L_l^{K_0}(\lambda)|_{X_r^{K_0}} \in \mathcal{L}(X_r^{K_0}, X_r^{K_0}).$$

We denote unstable dimensions $E_l^r(\lambda)$ and net crossing numbers $\chi_l^r$ as before

$$E_l^r(\lambda) : \text{the number of eigenvalues of } L_l^r(\lambda) \text{ with positive real part, counting algebraic multiplicity,} \tag{2.23}$$

$$\chi_l^r := \lim_{\lambda\to+\infty} \frac{1}{2}(E_l^r(\lambda) - E_l^r(-\lambda)). \tag{2.24}$$

Note that $E_l^0(\lambda)$ is defined consistently with (2.6) since $X^G = X_0^{H_0}$, in the setting of §2.1. The **net crossing number** $\chi_l^r$ counts how many conjugate complex pairs of eigenvalues of $L_l^r(\lambda)$ cross the imaginary axis from left to right as $\lambda$ sweeps through $\mathbb{R}$. By assumptions (2.20.$a,b$), $\chi_l^r$ is a well-defined finite integer.

Comparing the net crossing numbers $\chi_l^r$ from (2.24) with the local crossing numbers $\chi^r(\lambda_0)$ from (2.7), a slight discrepancy arises which involves real versus complex representations. Suppose we are in a generic case, i.e. only generic $G$-centers, $G = H_0/K_0$, occur in $X^{K_0}$ along our branches $(\lambda, x^l(\lambda))$. Then

$$\chi_l^r = \sum_{\text{branch } l} \chi^r(\lambda_0) + \sum_{\text{branch } l} \chi^{-r}(\lambda_0) \tag{2.25}$$

where the summation runs over all centers $(\lambda_0, x^l(\lambda_0))$ on the branch $(\lambda, x^l(\lambda))$. The reason is that we had to take real representations for the net crossing number $\chi_l^r$, while for the local crossing numbers $\chi^r(\lambda_0)$ a natural complexification was selected by the linear flow on the purely imaginary eigenspace, cf. (2.4.$a,b$). This canonical complexification allowed us to distinguish between $\chi^r$ and $\chi^{-r}$, locally, in contrast to the nongeneric case.

We can now define the global equivariant Hopf index for nongeneric $f$.

**2.8 Definition :**

*Let $f \in C^1$ be $\Gamma$-equivariant (cf. (1.6.a,b)), $H_0/K_0 \cong \mathbb{Z}_n$ a cyclic factor of $\Gamma$ (cf. (2.1)), and assume that (2.20.$a,b$) hold. Choose any binary orbit $d \in D(n)$ (cf. definition 2.4 and table 2.2). Then the* **global equivariant Hopf index** $\mathcal{H}_{H_0,K_0}^{\pm d}$ *is defined as*

$$\mathcal{H}_{H_0,K_0}^{\pm d} = \sum_{l=0}^{\bar{l}} \sum_{r\in d\cup(-d)} (-1)^{E_l^0(\lambda_0)} \chi_l^r \tag{2.26}$$

*where $E_l^0(\lambda)$, $\chi_l^r$ are defined as in (2.23), (2.24), and $(-1)^{E_l^0(\lambda_0)}$ is independent of the choice of $\lambda_0 \in \Lambda = \mathbb{R}$ by nondegeneracy assumption (2.20.$a$).*

In the setting of §2.1, with $H_0/K_0 \cong G = \mathbb{Z}_n$, the generic index $\mathcal{H}_n^d$ (definition 2.5) on $X^{K_0}$ relates to the nongeneric index $\mathcal{H}_{H_0,K_0}^{\pm d}$ (definition 2.8) by

$$\mathcal{H}_{H_0,K_0}^{\pm d} = \mathcal{H}_n^d + \mathcal{H}_n^{-d}, \tag{2.27}$$

due to (2.25), if we approximate $f$ by a generic $G$-equivariant nonlinearity on $X^{K_0}$. Such a definition by generic approximation, which is commonplace e.g. in Brouwer degree theory [Dei, Chow&Ha] and which relates to the question of homotopy invariance of the index $\mathcal{H}$, is discussed more in-depth in §9.4. We can now state our main results for nongeneric $f$, at last, recurring to the terminology of definitions 1.1, 1.2, 2.1, 2.4 and 2.8.

**2.9 Theorem :**
*Let $f \in C^1$ be $\Gamma$-equivariant as in (1.1), (1.6.a,b), $H_0/K_0 \cong \mathbb{Z}_\infty = \mathbb{R}/\mathbb{Z}$ a cyclic factor of $\Gamma$ as in (2.1) and assume that (2.20.a,b) hold. Choose any nonzero binary orbit $d \in D(\infty) \setminus \{0\}$. Finally, assume that the corresponding global equivariant Hopf index $\mathcal{H}^{\pm d}_{H_0,K_0}$, defined in (2.26), satisfies*

$$\mathcal{H}^{\pm d}_{H_0,K_0} \neq 0. \tag{2.28}$$

*Then there exists an* **unbounded continuum** *$\mathcal{C} \subseteq \Lambda \times X^{K_0}$ consisting of $H_0$-centers, of rotating waves, and possibly of frozen waves, with virtual symmetry at least*

$$(H_0, K, \Theta), \quad \text{and} \quad \Theta \in d \cup (-d), \; K_0 \leq K \leq H_0, \tag{2.29}$$

*for each element of $\mathcal{C}$.*

**2.10 Theorem :**
*Let $f \in C^1$ be $\Gamma$-equivariant as in (1.1), (1.6.a,b), $H_0/K_0 \cong \mathbb{Z}_n$ a finite or infinite cyclic factor of $\Gamma$ as in (2.1), and assume that (2.20.a-c) hold. Choose any binary orbit $d \in D(n)$. Finally, assume that*

$$\mathcal{H}^{\pm d}_{H_0,K_0} \neq 0. \tag{2.28}$$

*Then there exists a* **global continuum** *$\mathcal{C} \subseteq \Lambda \times X^{K_0}$ consisting of $H_0$-centers and of periodic solutions, with virtual symmetry at least*

$$(H_0, K, \Theta), \quad \text{and} \quad \Theta \in d \cup (-d), \; K_0 \leq K \leq H_0, \tag{2.29}$$

*for each element of $\mathcal{C}$. The continuum $\mathcal{C}$ contains both (uncountably many) periodic solutions and (at least) one center on one of the stationary branches $(\lambda, x^l(\lambda))$.*

*Here* **global** *means that*

$$\mathcal{C} \text{ is unbounded, or} \tag{2.30.a}$$

$$\mathcal{C} \text{ contains periodic solutions with arbitrarily large virtual periods.} \tag{2.30.b}$$

## 2.6 Variants

We briefly discuss some variants of the preceding results which are useful in applications, see §8.

**2.11 Remark :**
Failing of nondegeneracy assumption (2.20.$a$) on the global stationary branches $(\lambda, x^l(\lambda))$ can be quite obnoxious in applications. Sometimes it is more convenient to replace $\Lambda \times X$ by an open subset $\mathcal{Y} \subseteq \Lambda \times X$, and work in $\mathcal{Y}$, $\mathcal{Y} \cap (\Lambda \times X^{K_0})$, etc., instead. Then theorems 2.9, 2.10 hold true if we replace the requirement for $\mathcal{C} \subseteq \Lambda \times X$ to be unbounded by

$$\begin{array}{l} \mathcal{C} \subseteq \mathcal{Y} \text{ is unbounded, or} \\ \text{the closure of } \mathcal{C} \text{ intersects the boundary of } \mathcal{Y}, \end{array} \tag{2.30.a)'$$

in theorem 2.10 and, analogously, in theorem 2.9. For example, this allows us to take

$$\mathcal{Y} \quad := \quad (\Lambda \times X) \setminus \{\text{all annoying stationary solutions}\}.$$

Or, if the set of $H_0$-centers is unbounded, contrary to assumption (2.20.$b$), we may take

$$\mathcal{Y} \quad = \quad (-c, c) \times X$$

and let $c$ become large. Or we may be interested in positive solutions $x > 0$ (component-wise) and define $\mathcal{Y}$ accordingly. Or we may restrict attention to $\lambda > 0$, etc., etc.

**2.12 Remark :**

In §8 we are going to apply theorems 2.9, 2.10 in an analytic semigroup setting

$$d_t x \quad = \quad A(\lambda)x + f(\lambda, x)\,, \tag{2.31}$$

cf. [Hen] for a technical background. Specifically, we assume the following.

$A(\lambda)$ is a $C^4$-family of sectorial operators on a real Hilbert space $X$, with dense domain $\mathcal{D}(A(\lambda))$ which is independent of $\lambda \in \mathbb{R}$, and with compact resolvent. Further assume that $X_\omega := \mathcal{D}(A(\lambda)^\omega)$, equipped with the graph norm of $A(\lambda)^\omega$, is independent of $\lambda$ for some $\omega \in [0, 1)$. (2.32.$a$)

Here differentiability is understood in the uniform operator topology $\mathcal{L}(\mathcal{D}(A(0)), X)$, where $\mathcal{D}(A(0))$ is equipped with the graph norm of $A(0)$. For $f$ we require

$$f \in C^4(\Lambda \times X_\omega, X). \tag{2.32.$b$}$$

The compact Lie group $\Gamma$ comes in via its orthogonal representation on $X$. We assume

$f$ and $A(\lambda)$ are equivariant with respect to $\Gamma$, i.e. (2.32.$c$)

$$\begin{aligned} f(\lambda, \gamma x) &= \gamma f(\lambda, x), \\ A(\lambda)\gamma x &= \gamma A(\lambda)x, \end{aligned}$$

for all $\gamma \in \Gamma$, $\lambda \in \Lambda$, and $x \in X_\omega$ resp. $x \in \mathcal{D}(A(0))$.

In case $H_0/K_0 \cong \mathbb{Z}_\infty = \mathbb{R}/\mathbb{Z}$, we require for the canonical representation $\rho$ of $H_0/K_0$ on $X^{K_0}$ in addition that

$$\mathcal{R} := \tfrac{d}{dh}\rho(h)|_{h=0} \;:\quad X_\omega \cap X^{K_0} \to X^{K_0} \tag{2.32.$d$}$$

is a bounded operator.

To express genericity in this framework, we define

$\mathcal{F}$: the set of $f \in C^4(\Lambda \times (X_\omega \cap X^{K_0}), X^{K_0})$ which satisfy

$$f(\lambda, \rho(h)x) \;=\; \rho(h)f(\lambda, x) \tag{2.33}$$

for all $h \in H_0/K_0$, $\lambda \in \Lambda$, and $x \in X_\omega \cap X^{K_0}$.

Again, we endow $\mathcal{F}$ with the (weak) topology of uniform convergence on bounded sets. In this topology, $\mathcal{F}$ is a Baire space: countable intersections of open dense sets are still dense; cf. e.g. [Di, theorem 12.16.1]. Thus it makes sense to speak of generic subsets of $\mathcal{F}$. By compactness of the resolvent, the local semiflow defined by (2.31) is compact for any small positive time. Indicating any necessary modifications, we will model our proofs of the "finite-dimensional" theorems 2.6, 2.7, 2.9, 2.10 to fit to this infinite-dimensional case. With the obvious modification

$$L^{K_0} := (A(\lambda) + D_x f(\lambda, x))|_{X^{K_0}} \tag{2.19$'$}$$

entering into the definitions 2.5 and 2.8 of the global equivariant Hopf index $\mathcal{H}$ we then arrive at

**2.13 Corollary :**
*Under the additional assumptions (2.32.a-d) , theorems 2.6, 2.7, 2.9, 2.10 still hold true for analytic semigroups (2.31).*

# 3 No symmetry – a survey

## 3.1 Outline

Most of this chapter is intentionally didactic in character. We explain our basic concepts of proof on a purely intuitive level and in the simplest possible case: that of no symmetry, $\Gamma = \{id\}$. For a technically complete treatment of this case see [All&M-P&Y1,2, Chow&M-P, Chow&M-P&Y2, Fie2, M-P&Y1,2]. There are no lemma-lemma-lemma-theorem strings in this chapter. Instead we refer the reader to the appropriate places in §§4-7 and 10 where the respective arguments are carried out in the symmetry context. We only give some precise definitions, for later reference.

This chapter is organized as follows. In §3.2 we give a very brief account of global stationary bifurcation in one parameter. We base this account on generic approximations, to relate it to our approach. In §3.3 we describe the generic secondary bifurcations for periodic solutions: turn and flip doubling. Introducing an orbit index $\Phi$ in the case of no symmetry we finally sketch a proof of the corresponding versions of theorem 2.7, in §3.4, and of theorem 2.10, in §3.5.

## 3.2 Global stationary bifurcation

The classical global bifurcation result for stationary solutions (no symmetry) is due to Rabinowitz [Rab]. We describe it in a special setting which is convenient for us. We want to solve

$$\mathfrak{z}(\lambda, x) = 0 \tag{3.1}$$

say for $\lambda \in \Lambda = \mathbf{R}$ and $x \in X = \mathbf{R}^N$, $\mathfrak{z} \in C^2(\Lambda \times X, X)$. Assume that there is a trivial branch of solutions

$$\mathfrak{z}(\lambda, 0) = 0, \tag{3.2}$$

and that, for $|\lambda|$ large, all eigenvalues of the linearization $L(\lambda) := D_x\mathfrak{z}(\lambda, 0)$ are nonzero. The Brouwer degree at $(\lambda, 0)$, [Chow&Ha, Dei], is then given by

$$deg(\lambda, 0) = sgn\ det\ L(\lambda) = (-1)^N \cdot (-1)^{E(\lambda)}, \tag{3.3}$$

where $E(\lambda)$ is the number of eigenvalues of $L(\lambda)$ with positive real part, similarly to our definition (2.23) of $E^r(\lambda)$. Let us finally assume that

$$deg(\lambda, 0) \neq deg(-\lambda, 0), \quad \text{for large } \lambda > 0. \tag{3.4}$$

In terms of crossing numbers (see (2.24), (2.26), (2.28)) this means that

$$\lim_{\lambda \to \infty} (E(\lambda) - E(-\lambda)) \quad \text{is odd.} \tag{3.5}$$

As in the proof of the Rabinowitz theorem [Rab, theorem 1.3] we may then conclude that a global, i.e. unbounded, continuum of zeros $(\lambda, x)$ of $\mathfrak{z}$ bifurcates from the trivial branch. This result also follows from Ize [Ize1, p.77].

Let us sketch an idea of proof, taken from [Chow&Ha, §5.7], which indicates the method we have in mind for the case of periodic solutions. As a start, let us assume that $\mathfrak{z}$ is "generic". Here, "genericity" will only mean that 0 is a regular value of $\mathfrak{z}$, i.e. $\mathfrak{z}(\lambda_0, x_0) = 0$ implies for the total derivative

$$D\mathfrak{z}(\lambda_0, x_0) \text{ is surjective.}$$

By Sard's theorem [Ab&Ro] "genericity" holds, if not for $\mathfrak{z}$ itself, then at least for some perturbations

$$\mathfrak{z}(\lambda, x) + \varepsilon_i \tag{3.6}$$

with arbitrarily small constant vectors $\varepsilon_i$. Note that the trivial branch may get perturbed, or even get disconnected, by such a perturbation.

In case $\mathfrak{z}$ is "generic", the solution set $\mathfrak{z}^{-1}(0) \subset \Lambda \times X$ is an embedded one-dimensional $C^2$-manifold, i.e. a locally finite union of $C^2$-curves

$$s \mapsto (\lambda(s), x(s))$$

which are parametrized by arclength. Let us assign names to points on such curves, rigorously.

**3.1 Definition :**

*We call a zero $(\lambda_0, x_0)$ of $\mathfrak{z}$* ***x-regular*** *if all eigenvalues of the linearization $D_x\mathfrak{z}(\lambda_0, x_0)$ of $\mathfrak{z}$ with respect to $x$ are nonzero.*

*We call a zero $(\lambda_0, x_0) = (\lambda(s_0), x(s_0))$ of $\mathfrak{z}$ a* ***turn***, *if the following conditions (3.7.a-c) hold.*

$$\begin{aligned} &D\mathfrak{z}(\lambda_0, x_0) \text{ is surjective, but} \\ &D_x\mathfrak{z}(\lambda_0, x_0) \text{ has a simple eigenvalue } \mu_0 = 0. \end{aligned} \tag{3.7.a}$$

*In particular, the local stationary branch $(\lambda(s), x(s))$ through $(\lambda_0, x_0) = (\lambda(s_0), x(s_0))$ satisfies $D_s\lambda(s_0) = 0$ and $D_s x(s_0) \neq 0$ is an eigenvector of $\mu_0 = 0$. Let $\mu(s)$ denote the local continuation of $\mu_0$ to an eigenvalue of $D_x\mathfrak{z}(\lambda(s), x(s))$. We require a transversality condition*

$$D_s\mu(s_0) \neq 0. \tag{3.7.b}$$

*This condition turns out to be equivalent to the curvature condition*

$$D_s^2\lambda(s_0) \neq 0. \tag{3.7.c}$$

*A turn is drawn in fig. 3.1 below. As a variant we define a* ***turn*** *$(\lambda_0, x_0)$* ***of fixed points*** *of a map $Z \in C^3(\Lambda \times X, X)$ for later reference. We just require $(\lambda_0, x_0)$ to be a turn for*

$$\mathfrak{z}(\lambda, x) := Z(\lambda, x) - x,$$

*and assume in addition*

$$\text{the spectrum of } D_x Z(\lambda_0, x_0) \text{ on the unit circle consists of only the simple eigenvalue 1.} \tag{3.7.a$'$}$$

Restricting our class of "generic" $\mathfrak{z}$ even further, one can conclude that for generic $\mathfrak{z}$ the solution set consists only of $x$-regular points and (isolated) turns. From the definition (3.3) of Brouwer degree *deg* it is obvious that

$$deg(\lambda(s), x(s)) \in \{\pm 1\}$$

stays invariant near $x$-regular points and, by (3.7.$b$), changes sign at a turn. In [Chow&Ha, §2.11] this observation is used to prove homotopy invariance of *deg* under the homotopy parameter $\lambda$.

We can now prove the Rabinowitz result, for generic $z$. Just orient the solution curves $(\lambda(s), x(s))$ at $x$-regular points such that

$$\lambda \cdot deg \qquad \text{increases} \tag{3.8}$$

along the oriented curves. By the above, we may extend this orientation consistently through turns. By assumption (3.4), the curve segments $(\lambda, 0)$ for $|\lambda|$ large have opposite orientations for $\lambda$ positive resp. $\lambda$ negative. Thus they belong to two different oriented curves. But neither of these curves can remain bounded as we follow it inwards from $|\lambda| = \infty$. Indeed, because they are embedded curves without boundary, they cannot pile up in any bounded region. Thus each of these curves extends globally, away from $\Lambda \times \{0\}$. Passing to a general, nongeneric limit with these curves, one can then conclude existence of a global continuum (not necessarily a curve) bifurcating from $\Lambda \times \{0\}$. This is the Rabinowitz result.

The "usual" proof uses homotopy invariance of Brouwer degree directly. In constrast, we introduce the generic (i.e. nondegenerate) version (3.3) of Brouwer degree above in an ad-hoc fashion to orient our generic curves. Simultaneously, we can conclude homotopy invariance of Brouwer degree from its homotopy invariance at turns, as is done in [Chow&Ha, §2.11]. Of course, homotopy invariance of Brouwer degree can be established by other means. For example we mention the algebraic topology approach [Do, Dug]. For periodic solutions we favor the "generic" approach because it allows us to select curves along which the minimal period jumps discontinuously, if we prefer. Alternatively one may also reconstruct the more topologically minded Fuller index for periodic orbits this way, see [Ful, Chow&M-P].

## 3.3 Generic local bifurcations

We begin to describe generic bifurcations of periodic solutions in the case of no symmetry. Recall that generic Hopf bifurcation was specified in (2.3.$a$-$c$) already; these conditions specialize verbatim to the case of no symmetry.

To treat secondary bifurcations of periodic solutions into other periodic solutions, we employ the usual concept of a Poincaré section with an associated Poincaré map, see e.g. [Ab&Mars, ch. 7.1]. Let $(\lambda_0, x_0)$ be on a periodic solution $x_0(t)$ of (1.1). Then the Poincaré section $S = \{x_0\} + S'$ is a local affine hyperplane through $x_0$ of codimension 1 in $X$, such that $\dot{x}_0$ points out of $S$ (e.g. $S := \{x_0\} + \langle \dot{x}_0 \rangle^{\perp}$). The Poincaré map $\Pi$ is defined on a sufficiently small neighborhood of $x_0$ in $S$: it maps $x$ into the point where the positive trajectory $x(t)$, $t > 0$, through $x$ first hits $S$ again. We write, a bit sloppily,

$$\Pi(\lambda_0, \cdot): \qquad S_{loc} \to S. \tag{3.9}$$

Obviously, bifurcation of periodic solutions $(\lambda, x(t))$ from $(\lambda_0, x_0(t))$ is equivalent to bifurcation of fixed points $x$ of some iterate

$$[\Pi(\lambda, \cdot)]^k$$

from $(\lambda_0, x_0)$. Such bifurcations are indicated by the linearization of $\Pi$. The eigenvalues of

$$D_x \Pi(\lambda_0, x_0)$$

are called (nontrivial) Floquet multipliers of $(\lambda_0, x_0)$.

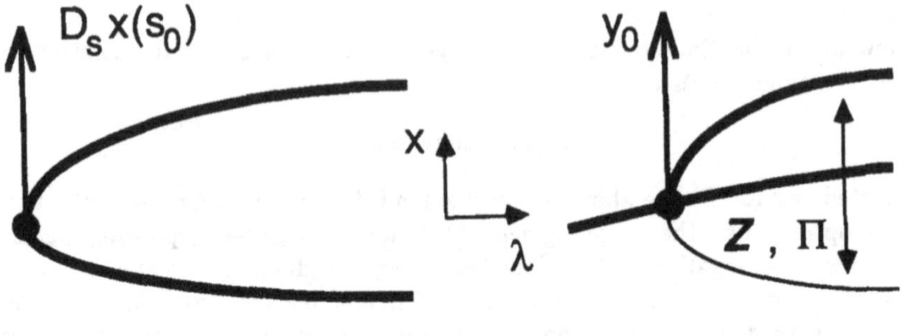


**Fig. 3.1** A turn

**Fig. 3.2** A flip

**3.2 Definition :** *Let $\Pi(\lambda_0, x_0) = x_0$. Then $(\lambda_0, x_0)$ is called* **type 0** *if no real or complex root of unity is a (nontrivial) Floquet multiplier. We call $(\lambda_0, x_0)$ a* **turn** *if it is a turn of the map $Z := \Pi$. We call $(\lambda_0, x_0)$ a* **flip doubling** *if it is a flip of the map $Z := \Pi$.*

*Replacing $S$ by $X$, here, $(\lambda_0, x_0)$ is called a* **flip** *of an abstract map $Z \in C^3(\Lambda \times X, X)$ if $Z(\lambda_0, x_0) = x_0$ and conditions (3.10.a-d) below hold. We do not require $Z$ to be a Poincaré map, for later convenience. In detail, we assume*

*$D_x Z(\lambda_0, x_0)$ has a simple eigenvalue $\mu(\lambda_0) = -1$, with eigenvector $y_0$, and this is the only eigenvalue on the unit circle.* (3.10.a)

*In particular, there is a unique local $C^3$-branch $x = x_\lambda$ of fixed points of $Z(\lambda, \cdot)$ through $x_0 = x_{\lambda_0}$. For the corresponding continuation $\mu(\lambda)$ of the eigenvalue $\mu_0 = \mu(\lambda_0) = -1$ we again impose the transversality condition*

$$D_\lambda \mu(\lambda_0) \neq 0. \tag{3.10.b}$$

*Then [Cra&Rab1] implies a pitchfork bifurcation of fixed points of $[Z(\lambda, \cdot)]^2$ at $(\lambda_0, x_0)$. The unique bifurcating local $C^2$-branch*

$$(\lambda(s), x(s)) \in \Lambda_{loc} \times S_{loc}, \quad |s| < \varepsilon$$

*can be $C^2$-parametrized such that*

$$\begin{aligned} &(\lambda(0), x(0)) = (\lambda_0, x_0) \\ &D_s\lambda(0) = 0, \;\; D_s x(0) = y_0 \\ &\lambda(s) = \lambda(-s), \;\; Z(\lambda(s), x(s)) = x(-s). \end{aligned} \tag{3.10.c}$$

*We finally require the curvature condition (which does not follow from (3.10.b), this time)*

$$D_s^2 \lambda(0) \neq 0. \tag{3.10.d}$$

*A flip is depicted in fig. 3.2.*

Poincaré [Poi, ch. XXXI] discusses the turn bifurcation in the context of periodically forced Hamiltonian systems with one degree of freedom. Note that a turn is called "type I" and a flip doubling is "type II" in [M-P&Y2]. Below we will encounter sufficiently many "types" to prefer a more descriptive terminology. For completeness we also note that a turn is called saddle-node bifurcation if one half of the branch is stable, see [Ab&Mars, Chow&Ha, Gu&Ho, Spa] and a flip doubling is called period doubling in [Gu&Ho, Spa] and also flip in [Gu&Ho].

The important thing about generic Hopf bifurcations, turns and flip doublings is that, for generic one-parameter $f$, they exhaust the zoo of bifurcations which involve only stationary and periodic solutions. All other periodic solutions are then of type 0. This result extends the Kupka-Smale theorem (the no-parameter case) and can be found in [Bru] for diffeomorphisms (like $\Pi$), in [All&M-P&Y2] and [Med] for ODEs, and in [Fie2] for analytic semigroups. Note that type 0 solutions are not necessarily hyperbolic, so bifurcations e.g. to invariant 2-tori may well occur. But they do not figure as bifurcations to **periodic** solutions, in our context. So these bifurcations are ignored. Including symmetry will enlarge our list of generic secondary bifurcations, because various symmetry-breaking effects have to be included. This list is developed in §5 and summarized in theorem 5.11; see also table 5.2.

It is rather cumbersome to establish this complete list of generic secondary bifurcations, even in the case of no symmetry. The basic idea, however, is simple. Linearizing Poincaré maps along a branch of periodic solutions we obtain generic one-parameter families, i.e. curves, of matrices. Given a complex root of unity $\varsigma \notin \{+1,-1\}$, we may certainly perturb this curve of matrices such that the associated curve of eigenvalues avoids $\varsigma$: the (real) codimension of $\varsigma$ in $\mathbb{C}$ is 2. But if $\varsigma \in \{+1,-1\}$, then we cannot avoid $\varsigma$ as an eigenvalue. Indeed, we cannot push a curve of (simple) real eigenvalues into the complex region, and the real codimension of $\varsigma$ in $\mathbb{R}$ is only 1. Still we may assume that eigenvalues, i.e. Floquet multipliers, cross $\varsigma$ transversely giving rise to turns or flip doublings, respectively. A technically correct implementation of this idea relies on transversality theory, developed e.g. in [Ab&Ro], which boils down to Sard's theorem. Complications arise already for the Kupka-Smale theorem: the application of transversality theory requires a reduction to periodic solutions of certain **minimal** periods. For equivariant problems, we defer this whole topic of transversality to §10.

## 3.4 Global generic Hopf bifurcation

With the list of generic local bifurcations at hand we would now like to detect a global branch of periodic solutions in the situation of theorem 2.7, but without symmetry, i.e. $G = \{0\}$, $n = 1$. For a reference see [M-P&Y1,2]. Similarly to the stationary case we would like to follow branches of periodic solutions, orienting them globally. Obviously we need some "degree", some "index": $\Phi$. But a problem immediately arises. At a flip doubling, three arcs of periodic solutions join together, see fig. 3.2. Following a global path of periodic solutions beyond a flip doubling we have to decide which way to go. The index $\Phi$ should tell us. But even worse, we might loop back onto a previously followed path at a flip doubling, as an example of Alligood & Mallet-Paret &Yorke [All&M-P&Y1] with $x \in \mathbb{R}^4$ shows, cf. fig. 3.3. The classical "resolution" of this difficulty is to consider global Hopf bifurcation as a two parameter problem, including period $p$ as an additional parameter [Ful, Ale&Y, Ize1]; for an equivalent setting see (1.23) and §1.3. Note that the "jug-handle", fig. 3.3, provides an unbounded continuum in $(p, \lambda, x)$ because the period $p$ is not required to be the minimal period. Indeed, suppose we enter the loop from the left with $p = p_0$ being the minimal period. Tracing out the loop counter-clockwise, the period $p$ gradually increases to $2p_0$ as we

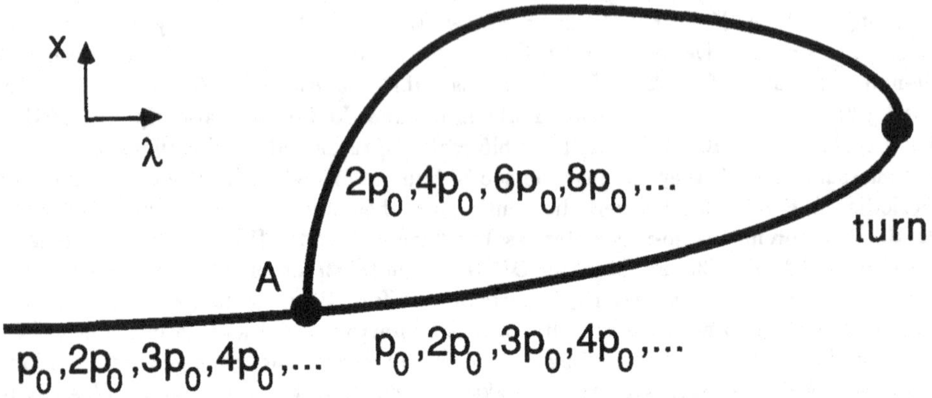


**Fig. 3.3** The "jug-handle" of Alligood & Mallet-Paret & Yorke [All&M-P&Y1].

return to the flip doubling point $A$. By continuity of $p$, we have to view $A$ now as a periodic solution with (non-minimal) period $2p_0$. Following the same loop repeatedly, we arrive at $A$ with (non-minimal) period $4p_0$, $8p_0$, and so on. Thus periods become arbitrarily large though the example looks perfectly bounded at first sight. Because primary and secondary periodic solutions near a flip doubling are linked in $\mathbb{R}^3$, the lowest possible dimension for such an example is $x \in \mathbb{R}^4$; but see also [Ale&Y2].

The orbit index $\Phi$, constructed by Mallet-Paret & Yorke [M-P&Y2], resolves the "jug-handle" difficulty as follows. Let $\sigma^+$ resp. $\sigma^-$ denote the number of real Floquet-multipliers of a type 0 periodic solution $(\lambda, x)$ in $(1, \infty)$ resp. $(-\infty, -1)$, counting algebraic multiplicities. Then

$$\Phi(\lambda, x) := \begin{cases} (-1)^{\sigma^+} & \text{if } \sigma^- \text{ is even,} \\ 0 & \text{if } \sigma^- \text{ is odd,} \end{cases} \tag{3.11}$$

is called the orbit index of $(\lambda, x)$. In other words,

$$\Phi \quad = \quad \frac{1}{2}\left((-1)^{\sigma^+} + (-1)^{\sigma^+ + \sigma^-}\right) \quad . \tag{3.11$'$}$$

The index $\Phi$ is in fact homotopy invariant. Choosing arcs with $\Phi \neq 0$ then tells us which way to go at a flip doubling. Indeed, $\sigma^-$ changes parity along the primary branch at a flip doubling $(\lambda_0, x_0)$, by definition 3.2, (3.10.$b$). Thus $\Phi \neq 0$ to one side of $(\lambda_0, x_0)$, and $\Phi = 0$ to the other side, on the primary branch. By homotopy invariance, $\Phi \neq 0$ on the secondary branch. Consequently, branches with $\Phi \neq 0$ extend to a unique path through $(\lambda_0, x_0)$.

Following branches with $\Phi \neq 0$, we will never enter a loop as in fig. 3.3, and get trapped. Indeed $\Phi \neq 0$ on the secondary branch, by the above. By definition 3.2, or directly by homotopy invariance, $\Phi$ changes sign at the turn but stays nonzero. Thus the whole loop consists of solutions with $\Phi \neq 0$ (except for the turn and the flip doubling). Consequently, $\Phi = 0$ on the remaining half of the primary branch and it is impossible to get trapped in the loop entering it with $\Phi \neq 0$ from outside.

For discrete cyclic symmetries, the orbit index $\Phi$ is introduced in definition 6.1. The proof of homotopy invariance, proposition 6.2, covers a substantial part of §6.

For global bifurcation of stationary solutions we assumed some odd net crossing (a change of degree) in (3.5). The analogous assumption for periodic solutions without symmetry is given by (2.14), (2.17)

$$\mathcal{H} \quad = \quad \sum_{centers} ⏀ \quad \neq \quad 0. \tag{3.12}$$

We omit the binary orbit $d = (0)$, $n = 1$, and $r = 0$ in this case. Recall definitions 2.2 and 2.3 of the center-index ⏀ of a generic center $(\lambda_0, x_0)$ with bifurcating local branch $(\lambda(s), x(s,\cdot))$ of periodic solutions. By exchange of stability, ⏀ of $(\lambda_0, x_0)$ and $\Phi$ of $(\lambda(s), x(s,\cdot))$ are related by

$$\Phi \quad = \quad ⏀ \cdot sgn(\lambda(s) - \lambda_0), \quad \text{for small } |s|, \tag{3.13}$$

cf. [M-P&Y2, fig. 7.1]. The symmetry analogue is proved in proposition 6.3.

With these preparations in mind we can now sketch a proof of our generic theorem 2.7 in the case of no symmetry $G = \{0\}$, $n = 1$. Analogously to the stationary case we may orient $C^0$-curves $(\lambda(s), x(s,\cdot))$ of periodic solutions of type 0 with $\Phi \neq 0$, globally, such that

$$\lambda \cdot \Phi \qquad \text{increases} \tag{3.14}$$

along the oriented curve. By homotopy invariance of $\Phi$, we may extend this orientation consistently through turns and flip doublings. Maximal oriented curves which begin or terminate at some center are called "snakes" in [M-P&Y2]. Let $S \subseteq \Lambda \times X$ denote a snake which is bounded together with its minimal periods. Then $S$ originates at some center $(\lambda_0, x_0)$ and, by maximality and boundedness, terminates at some other center $(\lambda_0', x_0')$ if we follow the orientation of $S$. By exchange of stability (3.13), the center-indices ⏀ resp. ⏀$'$ at $(\lambda_0, x_0)$ resp. $(\lambda_0', x_0')$ satisfy

$$⏀ = +1, \quad ⏀' = -1. \tag{3.15}$$

Note that ⏀ + ⏀$' = 0$. Conversely, because $\mathcal{H} = \sum$ ⏀ $\neq 0$ by assumption (3.12), there exists a snake $Z$ which is unbounded or contains arbitrarily large minimal periods. This proves theorem 2.7 in the case of no symmetry. With all the ingredients at hand, the proof of theorem 2.7 in the case $G = \mathbb{R}/\mathbb{Z}$ will actually be quite similar, see §6.6.

## 3.5 Global nongeneric Hopf bifurcation

It remains to comment on the general nongeneric result, theorem 2.10, in the case of no symmetry $\Gamma = H_0 = K_0 = \{id\}$. Brushing subtleties aside, for a moment, we approximate the nonlinearity $f$ by generic nonlinearities

$$g_i \to f. \tag{3.16}$$

By assumption (2.28) we know $\mathcal{H} \neq 0$ for $f$, which implies $\mathcal{H} \neq 0$ for the $g_i$. This yields global snakes $Z_i$ for $g_i$ and we may put

$$C \quad := \quad \text{"}\lim_i\text{"}\, Z_i \tag{3.17}$$

to obtain a global continuum $C$ of periodic solutions for $f$, as required in theorem 2.10.

Subtleties arise as follows. We have to make precise the "lim" in (3.17) and ensure $C$ is actually a continuum. This will be based on lemma 7.1 below, following [Why]. Then $C$ will indeed consist of (stationary and) periodic solutions for $f$. But note that the minimal period of the limit might be only a fraction of the limit of minimal periods. In particular: why should $C$ be global with respect to periods, even if the $Z_i$ are? This is the place where the notion of virtual period comes in. Indeed, the limit of minimal periods is still a virtual period of the limit [M-P&Y2, Chow&M-P&Y2]. Thus we will have arbitrarily large virtual periods of $C$ if the $Z_i$ stay uniformly bounded and have arbitrarily large minimal periods. On the other hand, if the $Z_i$ do not stay uniformly bounded then $C$ will be unbounded. In any case, $C$ is then global. Is it, really? In principle, the $Z_i$ could collapse to a continuum $C$ which consists only of centers while virtual periods on $C$ are unbounded. As in [Fie2], it requires analyticity assumption (2.20.c) and a quite careful construction of the $Z_i$, including continua $Z_i^! \subseteq Z$ of uniformly large virtual periods, to overcome this last subtlety.

Including symmetry complicates the situation by the notion of virtual symmetry. The details fill §4. Approximation problems are covered in §7. Note however that the appearance of rotating and frozen waves in theorems 2.6 and 2.9 is special to $SO(2)$-symmetry and has no counterpart in systems without symmetry. Curiously, these rotating/frozen waves appear as a hybrid between the global results for stationary resp. for periodic solutions discussed above. Indeed turns and flip doublings occur, generically (theorem 5.11). But we may extract bifurcating continua (of rotating/frozen waves) which are **unbounded** in $\Lambda \times X^{K_0}$, without caring about minimal or virtual periods too much. This is reminiscent of the global one-parameter result of Rabinowitz. With this apparent synthesis we conclude our didactic excursion into a world without symmetry.

# 4 Virtual symmetry

## 4.1 Outline

The notion of virtual symmetry, which we discuss in this section, is central to the idea of proving global results by (generic) approximation $f_i \to f$. Any approximation faces the following problem. Let $\xi_i(\cdot)$ be a sequence of periodic solutions, with minimal period $p_i > 0$ rescaled to 1 and with symmetry $(H_\infty, K_\infty, \Theta_\infty)$ independent of $i$. The $\xi_i$ thus satisfy

$$F(f_i, \lambda_i, \xi_i, p_i) \;:=\; -\frac{1}{p_i}\dot{\xi}_i + f_i(\lambda_i, \xi_i) \;=\; 0\,. \tag{4.1}$$

Assume $f_i \to f$, $\lambda_i \to \lambda$, $\xi_i \to \xi$, $p_i \to p_\infty$ in the appropriate spaces. Then $\xi$ is itself a periodic or stationary solution, but its minimal period $p$ need not be $p_\infty$, and even if $p = p_\infty$ then the symmetry $(H, K, \Theta)$ need not be given by $(H_\infty, K_\infty, \Theta_\infty)$. In general, $p_\infty$ is only some multiple of $p$, and even if $p_\infty = p$ then $\Sigma_\infty := H_\infty^{\Theta_\infty}$ is only a subgroup of $\Sigma := H^\Theta \leq \Gamma \times S^1$. In other words, minimal period and symmetry are not stable under limits.

In §3.5 we have noted that virtual periods were introduced by Chow, Mallet-Paret, and Yorke [M-P&Y2, Chow&M-P&Y2] to remedy this problem, as far as minimal periods are concerned. Below, we employ the notion of virtual symmetry to take care of the symmetry aspect as well. In fact, it turns out that the whole idea of virtual period can be subsumed under the symmetry point of view with no additional effort. In §4.2 we give a general definition of virtual isotropy which looks rather "stationary", and then state stability under limits in proposition 4.3. In §4.3 we observe that virtual symmetries and virtual periods are a special case of virtual isotropies. As a consequence virtual symmetries and virtual periods for periodic solutions are well-behaved under limits, cf. corollary 4.3. We conclude this section with three lemmas which involve virtual periods and virtual symmetries and which are stated for later reference.

## 4.2 Virtual isotropy

We develop our concept of virtual isotropy in reasonable generality although we later apply it only in the special setting (4.1). Therefore we deviate from the established notation to formulate definition 4.1 and proposition 4.3. Let $\Xi$, $\Xi'$ be Banach spaces with continuous actions of the group $\tilde{\Gamma}$ on $\Xi$ resp. $\Xi'$. Here $\Xi'$ does **not** denote the dual space of $\Xi$. Consider a $\tilde{\Gamma}$-equivariant $C^1$-map

$$F: \quad \Xi \to \Xi' \tag{4.2}$$

such that

$$F(\xi) \quad = \quad 0 \tag{4.3.a}$$

for some $\xi \in \Xi$, and such that $DF(\xi)$ is Fredholm [Kato].

**4.1 Definition :**

*Let $F(\xi) = 0$ as above. A subgroup $\Sigma$ of $\tilde{\Gamma}$ is called* **a virtual isotropy** *of $\xi$ with respect to $F$ if there exist elements $y^1, y^2, \cdots$ of $\Xi$ such that*

$$DF(\xi)\, y^j \quad = \quad 0, \qquad \text{for all } j, \text{ and} \tag{4.3.b}$$

$$\tilde{\Gamma}_{(\xi, y^1, y^2, \cdots)} \quad = \quad \Sigma \,. \tag{4.3.c}$$

*Here* $\tilde{\Gamma}_{(\xi,y^1,y^2,\cdots)} := \tilde{\Gamma}_\xi \cap (\bigcap_{j\geq 1} \tilde{\Gamma}_{y^j})$ *denotes the isotropy of* $(\xi, y^1, y^2, \cdots)$ *with respect to the obvious diagonal action of* $\tilde{\Gamma}$.

**4.2 Remark :**

Certainly $\Sigma \leq \tilde{\Gamma}_\xi$, for any virtual isotropy $\Sigma$. Picking $y^j = 0$ for all $j$ we see that the isotropy $\tilde{\Gamma}_\xi$ of $\xi$ is always a virtual isotropy of $\xi$. But in general, it may depend on $F$ whether some given $\Sigma \leq \tilde{\Gamma}_\xi$ is called a virtual symmetry or not. If $\tilde{\Gamma}_\xi$ acts trivially on $ker\ DF(\xi)$, then $\Sigma = \tilde{\Gamma}_\xi$ is the only virtual isotropy of $\xi$. This holds in particular if $ker\ DF(\xi) = \{0\}$. If, on the other hand, $\tilde{\Gamma}_\xi$ does not leave each element of $ker\ DF(\xi)$ fixed then other virtual isotropies besides $\tilde{\Gamma}_\xi$ occur as well. Therefore we have to keep $F$ in mind whenever we speak of virtual isotropy. But for brevity, we will not mention $F$ explicitly whenever the context is definitive.

Infinitely many $y^j$ are used in our definition above. Because $DF(\xi)$ is Fredholm, finitely many would actually be sufficient: we may put $y^j = 0$ for $j > dim\ ker\ DF(\xi)$. Indeed, we may renumber the $y^j$ such that

$$span\{y^j \mid 1 \leq j < \infty\} \quad = \quad span\{y^j \mid 1 \leq j \leq dim\ ker\ DF(\xi)\} \,.$$

But then

$$\bigcap_{j=1}^{\infty} \tilde{\Gamma}_{y^j} \quad = \quad \bigcap_{j=1}^{dim\ ker\ DF(\xi)} \tilde{\Gamma}_{y^j} \,,$$

proving our claim.

**4.3 Proposition :**

*Let* $\tilde{\Gamma}$ *be a compact Lie group, and* $F_i$ *a sequence of* $\tilde{\Gamma}$*-equivariant* $C^1$*-maps from* $\Xi$ *to* $\Xi'$ *with* $F_i(\xi_i) = 0$. *Assume that*

$$\xi_i \rightarrow \xi_\infty \,,$$
$$F_i \rightarrow F_\infty \,,$$

$C^1$*-uniformly on some closed neighborhood* $\mathfrak{V}$ *of* $\xi_\infty$ *and that*

$$DF_\infty(\xi_\infty)$$

*is a linear Fredholm operator [Kato]. Finally assume*

*the fixed subgroup* $\Sigma_\infty$ *of* $\tilde{\Gamma}$ *is a virtual isotropy of* $\xi_i$ *with respect to* $F_i$, *for each* $i$. (4.4)

*Then* $\Sigma_\infty$ *is a virtual isotropy of* $\xi_\infty$ *with respect to* $F_\infty$.

**Proof :**

The proof consists of a careful application of standard Ljapunov-Schmidt reduction, and it proceeds along the lines of [Pos1, theorem 3.2,1]. There, the special case $\tilde{\Gamma}_{\xi_\infty} = \tilde{\Gamma}$, $\Sigma_\infty = \tilde{\Gamma}_{\xi_i}$ was considered in a slightly different technical setting. We are indebted to Christoph Pospiech for suggesting this approach.

Let $P, Q$ denote projections onto $ker\ DF_\infty(\xi_\infty)$, $im\ DF_\infty(\xi_\infty)$, respectively. Because $\tilde{\Gamma}$ is assumed compact, $P$ and $Q$ can be taken $\tilde{\Gamma}_{\xi_\infty}$-equivariant (cf. e.g. [Van1]). Choose sequences $y_i^1, y_i^2, \ldots \in ker\ DF(\xi_i)$ such that

$$\tilde{\Gamma}_{(\xi_i, y_i^1, y_i^2, \ldots)} = \Sigma_\infty .$$

To simplify notation define

$$\xi_i^1 := P(\xi_i - \xi_\infty), \qquad \xi_i^2 := (id - P)(\xi_i - \xi_\infty).$$

The following two claims will be proved below for large enough $i$:

$$\tilde{\Gamma}_{(\xi_\infty, \xi_i^1)} = \tilde{\Gamma}_{\xi_\infty} \cap \tilde{\Gamma}_{\xi_i} \tag{4.5}$$

$$\tilde{\Gamma}_{\xi_i} \cap \tilde{\Gamma}_{P y_i^j} = \tilde{\Gamma}_{\xi_i} \cap \tilde{\Gamma}_{y_i^j} \quad , \qquad \text{for all } j. \tag{4.6}$$

Using (4.5) and (4.6), the proof can be completed as follows. Obviously, $\tilde{\Gamma}_{\xi_\infty} \geq \Sigma_\infty$ by continuity and because $\tilde{\Gamma}_{\xi_i} \geq \Sigma_\infty$. This implies

$$\begin{aligned}
\tilde{\Gamma}_{(\xi_\infty, \xi_i^1, P y_i^1, P y_i^2, \ldots)} &= \tilde{\Gamma}_{(\xi_\infty, \xi_i^1)} \cap \left( \bigcap_{j \geq 1} \tilde{\Gamma}_{P y_i^j} \right) = \\
&= \tilde{\Gamma}_{\xi_\infty} \cap \left( \bigcap_{j \geq 1} (\tilde{\Gamma}_{\xi_i} \cap \tilde{\Gamma}_{P y_i^j}) \right) = \\
&= \tilde{\Gamma}_{\xi_\infty} \cap \tilde{\Gamma}_{\xi_i} \cap \left( \bigcap_{j \geq 1} \tilde{\Gamma}_{y_i^j} \right) = \\
&= \tilde{\Gamma}_{\xi_\infty} \cap \Sigma_\infty = \Sigma_\infty .
\end{aligned} \tag{4.7}$$

By construction, $\xi_i^1$ and all $P y_i^j$ are in $ker\ DF_\infty(\xi_\infty)$. Thus $\Sigma_\infty$ is a virtual isotropy of $\xi_\infty$, and it remains only to prove (4.5) and (4.6).

To prove (4.5) we define the evaluation map

$$\begin{aligned}
\Psi : \quad C^1(\mathcal{V}, \Xi) \times \mathcal{V}_0 \quad &\rightarrow \quad Q\Xi' \\
\Psi(F, \xi^1, \xi^2) \quad &:= \quad QF(\xi_\infty + \xi^1 + \xi^2)
\end{aligned}$$

where $\xi^1 := P(\xi - \xi_\infty)$, $\xi^2 := (id - P)(\xi - \xi_\infty)$, and $\mathcal{V}_0 := \mathcal{V} - \xi_\infty$ is $\mathcal{V}$ shifted to the origin. By construction, $\Psi$ is $C^1$, $\tilde{\Gamma}_{\xi_\infty}$-equivariant in $(\xi^1, \xi^2)$, $\Psi(F_\infty, 0, 0) = 0$, and

$$D_{\xi^2}\Psi(F_\infty, 0, 0) : \qquad (id - P)\Xi \quad \rightarrow \quad Q\Xi'$$

is an isomorphism. By the implicit function theorem, any solution of

$$\Psi(F, \xi^1, \xi^2) = 0$$

has the form

$$\xi^2 = \xi^2(F, \xi^1),$$

for $F$ near $F_\infty$, if $\tilde{\mathcal{V}}$ is chosen small enough. By uniqueness, $\xi^2(F, \cdot)$ is $\tilde{\Gamma}_{\xi_\infty}$-equivariant. Because $\xi_i^2 = \xi^2(F_i, \xi_i^1)$ we thus obtain

$$\tilde{\Gamma}_{\xi_\infty} \cap \tilde{\Gamma}_{\xi_i} = \tilde{\Gamma}_{\xi_\infty} \cap \tilde{\Gamma}_{\xi_i^1 + \xi_i^2} = \tilde{\Gamma}_{\xi_\infty} \cap \tilde{\Gamma}_{\xi_i^1} = \tilde{\Gamma}_{(\xi_\infty, \xi_i^1)}$$

and (4.5) is proved.

To prove (4.6) we note that $DF_i(\xi_i) y_i^j = 0$ and hence

$$QDF_i(\xi_i)(id - P) y_i^j = -QDF_i(\xi_i) P y_i^j .$$

This equation can be solved for $(id - P) y_i^j$. Indeed

$$QDF_i(\xi_i): \quad (id - P)\Xi \rightarrow Q\Xi',$$

being a small perturbation of $QDF_\infty(\xi_\infty)$, is an isomorphism for large enough $i$. Thus $(id - P) y_i^j$ depends $\tilde{\Gamma}_{\xi_i}$-equivariantly on $P y_i^j$ and (4.6) follows as (4.5) did. This completes the proof of proposition 4.3. □

**4.4 Remark :**

Below we will apply proposition 4.3 in the special case that

$$\Sigma_\infty = \tilde{\Gamma}_{\xi_i}$$

is the isotropy of $\xi_i$ itself. Putting $y_i^j = 0$ for all $i, j$ in the proof of proposition 4.3, and in particular in (4.7), it is then clear that for $i$ large enough

$$\Sigma_\infty = \tilde{\Gamma}_{(\xi_\infty, \xi_i^1)}$$

is a virtual isotropy of $\xi_\infty$, with $y^1 := \xi_i^1$ and all other $y^j = 0$. Thus a single element of $ker\ DF(\xi_\infty)$ is sufficient to represent the virtual symmetry $\Sigma_\infty$.

Following [Fie4] we arrive at global Hopf bifurcation results for integral equations via two successive approximations: first a generic approximation leading to a nongeneric ODE-result (this step is our main concern here), and then an ODE-approximation leading to the integral equation. It seems impossible to achieve this by a single approximation step, say by passing to a diagonal sequence. According to the proposition given above, we may then need two elements of $ker\ DF(\xi_0)$ to represent the virtual symmetries on the second (integral equation) level. It is not clear at present whether there is a more natural concept of virtual isotropy, stable under limits, which would involve only $\xi_\infty$ and a single element $y^1$ of $ker\ DF_\infty(\xi_\infty)$.

**4.5 Remark :**

Compactness of the group $\tilde{\Gamma}$ is used only in proposition 4.3. It enters into equivariance of the projections $P$ and $Q$. Indeed projections can be made equivariant using the Haar measure, see [Van1, theorem 2.5.9]. If $\Xi$ and $\Xi'$ are both Hilbert spaces with orthogonal action of $\tilde{\Gamma}$, we may drop the assumption that $\tilde{\Gamma}$ is compact. Just take orthogonal projections for $P$ and $Q$.

## 4.3 Virtual symmetry

After these generalities on virtual isotropy we return to the specific problem of periodic and stationary solutions in the setting (4.1). Recall our definition 1.2 of virtual period and virtual symmetry: $x(t)$ has virtual period $q > 0$ and virtual symmetry $(\hat{H}, \hat{K}, \hat{\Theta})$ if there exists a solution $y(t)$ of the linearized equation such that the pair $(x, y)$ has minimal period $q$ and symmetry $(\hat{H}, \hat{K}, \hat{\Theta})$. From proposition 4.3 we derive

**4.6 Corollary :**

*Let $\Gamma$ be a compact Lie group and $f_i$ a sequence in $C^1(\Lambda \times X,\ X)$, $\Gamma$-equivariant in $x \in X$, with nonstationary periodic solutions $(\lambda_i, x_i(\cdot))$ of minimal period $p_i > 0$. Assume that*

$$\begin{array}{lll} \lambda_i & \to \lambda_\infty & \text{in } \Lambda, \\ p_i & \to p_\infty & \text{in } \mathbb{R}, \\ x_i(\cdot) & \to x_\infty(\cdot) & \text{in } C^0(\mathbb{R}, X), \text{ and} \\ f_i & \to f_\infty, & \end{array}$$

*$C^1$-uniformly on bounded subsets of $X$. Finally assume*

$$(\hat{H}, \hat{K}, \hat{\Theta}) \text{ is the symmetry of } x_i(\cdot), \text{ for all } i. \tag{4.8}$$

*Then $(\hat{H}, \hat{K}, \hat{\Theta})$ is a virtual symmetry and $p_\infty$ is a virtual period of $x_\infty(\cdot)$.*

*Likewise, let $(\lambda_i, x_i)$ be on a rotating or a frozen wave of $f_i$ with symmetry $(\hat{H}, \hat{K}, \hat{\Theta})$ or, resp., $(\hat{H}, \hat{K}, \pm\hat{\Theta})$, independent of $i$. Then $(\lambda_i, x_i)$ satisfies*

$$-\alpha_i R x_i + f_i(\lambda_i, x_i) \quad = \quad 0, \tag{1.21$'$}$$

*as in §1.2. Assume*

$$\begin{array}{lll} \lambda_i & \to \lambda_\infty & \text{in } \Lambda, \\ \alpha_i & \to \alpha_\infty & \text{in } \mathbb{R}, \\ x_i(\cdot) & \to x_\infty & \text{in } X, \quad \text{and} \\ f_i & \to f_\infty, & \end{array}$$

*$C^1$-uniformly on bounded subsets of $X$.*

*Then $x_\infty$ is*

$$\text{on a rotating wave, or} \tag{4.9.a}$$

$$\text{on a frozen wave, or} \tag{4.9.b}$$

$$\text{a center in } X^{\hat{H}}, \text{ or} \tag{4.9.c}$$

$$\text{a stationary solution in } X^{\hat{H}} \text{ with} \tag{4.9.d}$$

$$\dim\left(X^{\hat{K}} \cap \ker D_x f(\lambda_\infty, x_\infty)\right) \geq 2.$$

*Moreover, $(\hat{H}, \hat{K}, \hat{\Theta})$ is a virtual symmetry of $x_\infty$ in cases (4.9.a, c) and $(\hat{H}, \hat{K}, \pm\hat{\Theta})$ is a virtual symmetry of $x_\infty$ in case (4.9.b).*

**Proof :**

Consider the periodic case first. We apply proposition 4.3 to the setting (4.1). Define

$$\begin{aligned} \xi_i(\tau) &:= x_i(p_i\tau) \\ F_i &:= F(f_i, \lambda_i, \cdot, p_i). \end{aligned}$$

Then $F_i$ is a $C^1$-map from $\Xi := \tilde{C}^1(\boldsymbol{R}, X)$ to $\Xi' := \tilde{C}^0(\boldsymbol{R}, X)$. In §1.2 we have already noticed equivariance of $F_i$ with respect to $\tilde{\Gamma} := \Gamma \times S^1$ acting by

$$\tilde{\rho}(\gamma, \vartheta)\ \xi_i(\cdot) \quad := \quad \rho(\gamma)\ \xi_i(\cdot - \vartheta). \tag{1.25}$$

Obviously our assumptions guarantee $F_i(\xi_i) = 0$, $\xi_i \to \xi_\infty$ in $\tilde{C}^0(\boldsymbol{R}, X)$ and $F_i \to F_\infty$, $C^1$-uniformly on bounded subsets of $\tilde{C}^1(\boldsymbol{R}, X)$, if we define

$$\begin{aligned} \xi_\infty(\tau) &:= x_\infty(p_\infty \tau) \\ F_\infty &:= F(f_\infty, \lambda_\infty, \cdot, p_\infty). \end{aligned} \tag{4.10}$$

Because $F_i(\xi_i) = 0$, (4.1) implies $\xi_i \to \xi_\infty$ even in $\tilde{C}^1(\boldsymbol{R}, X)$. Further note that $F_\infty$ is Fredholm [Van1, §2.2]. Finally,

$$\Sigma_\infty \quad := \quad \hat{H}^{\hat{\theta}} \quad = \quad \{(\hat{h}, \hat{\Theta}(\hat{h})) \mid \hat{h} \in \hat{H}\}$$

is the isotropy of $\xi_i$, for each $i$, as was observed in (1.26). Thus $\Sigma_\infty$ is a virtual isotropy of $\xi_\infty$. By remark 4.4, a single $y^1 \in ker\ DF_\infty(\xi_\infty)$ suffices to represent this virtual symmetry:

$$\hat{H}^{\hat{\Theta}} \quad = \quad \tilde{\Gamma}_{(\xi_\infty, y^1)}\ .$$

It remains to reinterpret this virtual isotropy on the $\xi_\infty$ - level as a virtual symmetry of the periodic solution $x_\infty$ . First note that $p_\infty > 0$. Otherwise $\xi_\infty$ and $y^1$ are both constants, by (4.1), and consequently

$$(id, \vartheta) \quad \in \quad \tilde{\Gamma}_{(\xi_\infty, y^1)} \quad = \quad \hat{H}^{\hat{\Theta}} \quad = \quad \tilde{\Gamma}_{\xi_i}$$

for some $\vartheta \in S^1 \setminus \{0\}$. This is a contradiction, because $\xi_i$ has minimal period 1. Thus $p_\infty$ is indeed positive. The same argument shows that the minimal period of the pair $(\xi_\infty, y^1)$ is 1. Rescaling this minimal period 1 by $y(t) := y^1(t/p_\infty)$, to become $p_\infty$, identifies $p_\infty$ as a virtual period and $(\hat{H}, \hat{K}, \hat{\Theta})$ as a virtual symmetry of $x_\infty$. This completes the proof of the periodic case.

Now consider the case of rotating and frozen waves. Passing to the limit in $(1.21)'$ we get

$$-\alpha_\infty \mathcal{R} x_\infty + f_\infty(\lambda_\infty, x_\infty) \quad = \quad 0$$

If $\mathcal{R} x_\infty \neq 0$, $\alpha_\infty \neq 0$, then $x_\infty$ is on a rotating wave:

$$x_\infty(t) \quad = \quad exp(\alpha_\infty \mathcal{R} t)\ x_\infty(0)\,.$$

Likewise, the $x_i$ are on rotating waves for $i$ large enough. In particular their minimal periods are uniformly bounded:

$$p_i \leq |1/\alpha_i|\,,$$

cf. (1.19). Passing to a subsequence, the $p_i$ may therefore be assumed to converge. Then the first part of corollary 4.6 shows that $(\hat{H}, \hat{K}, \hat{\Theta})$ is indeed a virtual symmetry of $x_\infty$, as required in (4.9.$a$).

If $\mathcal{R}x_\infty \neq 0$, $\alpha_\infty = 0$, then $x_\infty$ is on a frozen wave. Transforming the systems $f_i$, $f_\infty$ with some fixed $\varepsilon > 0$ to

$$\begin{aligned} y(t) &:= exp\,(-\varepsilon \mathcal{R} t) x(t) \\ \hat{f}_i(\lambda, y) &:= -\varepsilon \mathcal{R} y + f_i(\lambda, y) \\ \hat{f}_\infty(\lambda, y) &:= -\varepsilon \mathcal{R} y + f_\infty(\lambda, y)\,, \end{aligned}$$

as in (1.20), (1.21), this case reduces to the previous one because $y_\infty$ becomes a rotating wave, cf. case (4.9.$b$).

If $\mathcal{R}x_\infty = 0$, then $f_\infty(\lambda_\infty, x_\infty) = 0$. Thus $x_\infty$ is a stationary solution in $X^{\hat{H}}$. We may apply proposition 4.3 directly to

$$F_i(x) \quad := \quad -\alpha_i \mathcal{R} x + f_i(\lambda_i, x)$$

with $x = \xi \in X = \Xi = \Xi'$, this time. Because $\hat{K}$ is the isotropy of each $x_i$, we conclude that $\hat{K}$ is the isotropy of a pair $(x_\infty, y_0)$ with $\mathcal{R}y_0 \neq 0$ and

$$-\alpha_\infty \mathcal{R} y_0 + D_x f_\infty(\lambda_\infty, x_\infty) y_0 \quad = \quad 0\,.$$

If $\alpha_\infty \neq 0$, then

$$y(t) \quad := \quad exp(\alpha_\infty \mathcal{R} t) y_0$$

is a rotating wave solution of the linearized equation

$$\dot{y} \quad = \quad D_x f_\infty(\lambda_\infty, x_\infty) y\,.$$

Thus $x_\infty$ is a center with virtual symmetry $(\hat{H}, \hat{K}, \hat{\Theta})$ cf. case (4.9.$c$). But if $\alpha_\infty = 0$, then $y_0$ and $\mathcal{R}y_0 \neq 0$ are linearly independent elements of $X^{\hat{K}} \cap ker\ D_x f_\infty(\lambda_\infty, x_\infty)$, as required in case (4.9.$d$). This completes the proof of corollary 4.6. □

**4.7 Remark :**

Our notion of virtual isotropy generalizes the notion of virtual period in the case of no symmetry $\Gamma = \{id\}$. In fact, $p > 0$ is a virtual period of $x$ if and only if $\Sigma := \{(id, 0)\} \leq \tilde{\Gamma} := \Gamma \times S^1$ is a virtual symmetry of

$$\xi(\tau) := x(p\tau)\,.$$

This follows directly from definitions 1.2, 4.1 and the following claim. Given $y^j \in ker\ DF(\xi)$ with

$$\tilde{\Gamma}_{(\xi, y^1, y^2, \ldots)} \quad = \quad \{(id, 0)\}\,, \tag{4.11.a}$$

there exists a single $y \in ker\ DF(\xi)$ such that

$$\tilde{\Gamma}_{(\xi, y)} \quad = \quad \{(id, 0)\}\,. \tag{4.11.b}$$

To prove this claim we may assume, by remark 4.2, that only finitely many $y^j$ are nonzero, say those with $j = 1, \cdots, \bar{j}$. Defining

$$y \;:=\; \sum_{j=1}^{\bar{j}} \varepsilon_j y^j ,$$

(4.11.$b$) holds for "most" $\varepsilon_j$. Indeed, suppose on the contrary that (4.11.$b$) does not hold. Then the minimal period of the pair $(\xi, y)$ is less than 1. Thus $1/k$ is a period of $(\xi, y)$, for some integer $k > 1$, i.e.

$$\xi(1/k) \;=\; \xi(0) \quad \text{and} \quad \sum_{j=1}^{\bar{j}} \varepsilon_j \left(y^j(1/k) - y^j(0)\right) \;=\; 0. \tag{4.12}$$

By (4.11.$a$), $1/k$ cannot be a period of $(y^1, \cdots, y^{\bar{j}})$. Thus the set of $(\varepsilon_1, \cdots, \varepsilon_{\bar{j}})$, for which (4.12) holds, defines a hyperplane in $\mathbb{R}^{\bar{j}}$ of codimension at least one. Varying $k \in \mathbb{N}$ we conclude that (4.11.$b$) holds for "most" $\varepsilon_j$, namely for all those $(\varepsilon_1, \cdots, \varepsilon_{\bar{j}})$ which do not lie on any of the countably many hyperplanes. This proves equivalence of (4.11.$a$) and (4.11.$b$). Thus virtual isotropy generalizes virtual period. In particular, all previous results [M-P&Y2, Chow&M-P&Y2, Fie4] on limits of minimal resp. virtual periods being virtual periods become a corollary to proposition 4.3.

In the next lemma we determine all virtual periods and virtual symmetries of a stationary center $(\lambda_0, x_0)$ in terms of its purely imaginary eigenvalues and of the group action on their eigenspaces. This recipe is included for completeness and because it exhibits an interesting structure. This time, we complexify the linearization $D_x f(\lambda_0, x_0)$ and the representation $\rho$ of $\Gamma$ to $X^c := X \oplus iX$ in the usual way. For $y \in X^c$ we denote by

$$\Gamma_{x_0}(\langle y\rangle^c) \;:=\; \{\gamma \in \Gamma_{x_0} \mid \rho(\gamma)\langle y\rangle^c \;=\; \langle y\rangle^c\}$$

the set of those $\gamma \in \Gamma_{x_0}$ which leave the space $\langle y\rangle^c$, the complex span of $y$, invariant. This should be contrasted with

$$\Gamma_{(x_0,y)} \;=\; \{\gamma \in \Gamma_{x_0} \mid \rho(\gamma)y \;=\; y\}.$$

For later reference, the lemma also extracts the virtual periods (but not the virtual symmetries) of a periodic solution. In view of corollary 4.6, examples for virtual symmetries are given by the many generic secondary bifurcations listed in §5, table 5.2.

**4.8 Lemma :**

*Let $(\lambda_0, x_0)$ be a stationary solution of (1.1) for which all eigenvalues of the linearization $D_x f(\lambda_0, x_0)$ are nonzero. We denote the mutually distinct purely imaginary eigenvalues, if any, by $i\beta_j$ with $1 \leq |j| \leq \bar{j}$ and with the numbering $0 < \beta_j = -\beta_{-j}$ for $j > 0$.*

*Then $(\lambda_0, x_0)$ has a virtual period $q > 0$ and a virtual symmetry $(\hat{H}, \hat{K}, \hat{\Theta})$ if and only if there exists a nonempty subset*

$$J \subseteq \{j \mid 1 \leq j \leq \bar{j}\}$$

*and eigenvectors $y_j \in (X^{\hat{K}})^c$ of $i\beta_j$, for $j \in J$, such that conditions (4.13.a-d) below all hold. We denote the symmetry of $(x_0, y_j)$ by $(H_j, K_j, \Theta_j)$. Obviously*

$$H_j \;=\; \Gamma_{x_0}(\langle y_j\rangle^c)$$
$$K_j \;=\; \Gamma_{(x_0,y_j)} \;.$$

*The conditions are the following:*

$$q \quad = \quad lcm\ \{2\pi/\beta_j \,|\, j \in J\}, \tag{4.13.a}$$

*where lcm denotes the least common multiple;*

*$\hat{H}$ is the set of all $h \in \bigcap_{j\in J} H_j$ for which the congruences*

$$\frac{q}{2\pi/\beta_j}\cdot \vartheta \quad \equiv \quad \Theta_j(h) \qquad (mod\ \mathbb{Z}) \tag{4.13.b}$$

*admit a simultaneous solution $\vartheta \in \mathbb{R}/\mathbb{Z}$, for all $j \in J$;*

$$\hat{\Theta}(h) \quad \equiv \quad \vartheta \qquad (mod\ \mathbb{Z}) \tag{4.13.c}$$

*if $\vartheta$ is such that (4.13.b) holds; and finally*

$$\hat{K} \quad = \quad \bigcap_{j\in J} K_j\,. \tag{4.13.d}$$

*In particular $(\lambda_0, x_0)$ is an $\hat{H}$-center because $\hat{H} \le \Gamma_{x_0}$ by (4.13.b). Moreover, $(\lambda_0, x_0)$ has at most $2^{\bar{j}}$ distinct virtual periods, by (4.13.a).*

*Now let $(\lambda_0, x_0)$ be a periodic solution of (1.1) with minimal period p. Then q is a virtual period of $(\lambda_0, x_0)$ if, and only if, $(\lambda_0, x_0)$ has primitive $e_j$-th roots of unity, $1 \le j \le \bar{j}$, as Floquet-multipliers and $q/p$ is the least common multiple of a subset of $\{e_j \,|\, 1 \le j \le \bar{j}\}$. In particular, the minimal period is always a virtual period.*

**Proof :**

First we consider a stationary solution $(\lambda_0, x_0)$ assuming that $(\lambda_0, x_0)$ has virtual period $q$ and virtual symmetry $(\hat{H}, \hat{K}, \hat{\Theta})$. Then there exists $y(t) \in X$ with minimal period $q > 0$ satisfying

$$\dot{y} \quad = \quad D_x f(\lambda_0, x_0) y \tag{4.14}$$

such that $(x_0, y)$ has symmetry $(\hat{H}, \hat{K}, \hat{\Theta})$. By $L^2$ Fourier decomposition we may write

$$y(t) \quad = \quad \sum_{k\in\mathbb{Z}} \tilde{y}_k e^{ik(2\pi/q)t}\,, \tag{4.15}$$

and $\tilde{y}_k \ne 0$ implies that

$$i\beta \quad = \quad ik \cdot 2\pi/q$$

is an eigenvalue of $D_x f(\lambda_0, x_0)$ with eigenvector $\tilde{y}_k$. Therefore we may rewrite (4.15) as

$$y(t) \quad = \quad \sum_{1\le |j| \le \bar{j}} y_j e^{i\beta_j t}\quad . \tag{4.16}$$

In particular, the summation in (4.16) is finite. Define

$$J \quad := \quad \{j > 0 \,|\, y_j \ne 0\}.$$

$J$ is nonempty because $y \not\equiv 0$. Noting that $q > 0$ is the minimal period of $y(\cdot)$, we obtain

$$q \quad = \quad lcm\{2\pi/\beta_j \,|\, j \in J\} \tag{4.13.a}$$

from (4.16). Note here that $y_{-j} = \bar{y}_j$, because $y$ is real.

To prove (4.13.$b,c$) let $j \in J$ be arbitrary and $y_j \neq 0$ as above. Note that

$$y_j = \frac{1}{q}\int_0^q y(t)\, e^{-i\beta_j t}\, dt,$$

by Fourier decomposition. Therefore, $\rho(h)y(t) = y(t + \vartheta q)$ holds for each $t \in \boldsymbol{R}$ if and only if

$$\rho(h)y_j = e^{i\beta_j\vartheta q} y_j, \qquad \text{for all } j \in J. \tag{4.17.a}$$

On the other hand we have for each $h \in \bigcap_{j\in J} H_j$ that

$$\rho(h)y_j = e^{2\pi i\Theta_j(h)} y_j, \qquad \text{for all } j \in J, \tag{4.17.b}$$

because $(H_j, K_j, \Theta_j)$ is the symmetry of $(x_0, y_j)$.

Now suppose $h \in \hat{H}$, $\vartheta := \hat{\Theta}(h)$. Then (4.17.$a$) implies $h \in \bigcap_{j\in J} \Gamma_{x_0}(\langle y_j\rangle^c) = \bigcap_{j\in J} H_j$. Thus (4.17.$b$) applies, too, and

$$\frac{q}{2\pi/\beta_j}\vartheta \equiv \Theta_j(h), \qquad (mod\ \mathbb{Z}),$$

for all $j \in J$.

Conversely suppose $h \in \bigcap_{j\in J} H_j$ and $\vartheta$ solves all the above congruences. Then (4.17.$b$) implies (4.17.$a$), and consequently $h \in \hat{H}$, $\hat{\Theta}(h) \equiv \vartheta$ $(mod\ \mathbb{Z})$. This proves (4.13.$b,c$).

From (4.13.$b,c$) we obtain

$$\hat{K} = ker\ \hat{\Theta} = \bigcap_{j\in J} ker\ \Theta_j = \bigcap_{j\in J} K_j .$$

This proves (4.13.$d$) and thus shows the one direction of the equivalence claimed for stationary solutions in the lemma.

To show the other direction we just define $y(t)$ by (4.16), putting $y_j = 0$ for $j \notin J$. Then $q$ is the minimal period of $y(t)$, by (4.13.$a$), and hence it is a virtual period of $x_0$. Defining $(\hat{H}, \hat{K}, \hat{\Theta})$ by (4.13.$b$-$d$), and $(\tilde{H}, \tilde{K}, \tilde{\Theta})$ to be the virtual symmetry of $(\lambda_0, x_0)$ associated to $y$, we see from the previous reasoning that $\tilde{H} = \hat{H}$, $\tilde{K} = \hat{K}$, and $\tilde{\Theta} = \hat{\Theta}$. Thus $(\lambda_0, x_0)$ has indeed a virtual symmetry $(\hat{H}, \hat{K}, \hat{\Theta})$.

The claims for the virtual period of a periodic solution $(\lambda_0, x_0)$ are proved by linear algebra similar to the stationary case, and we omit the details which are given e.g. in [Fie2]. This finishes the proof. □

**4.9 Remark :**

The previous lemma helps us to conceptually separate the geometric aspect from the dynamic aspect. Geometrically, we are given representations on eigenspaces, eigenvectors $y_j$, and their isotropies $H_j = \Gamma_{x_0}(\langle y_j\rangle^c)$, $K_j = \Gamma_{(x_0,y_j)}$ . Even the symmetry $(H_j, K_j, \Theta_j)$ of the pair $(x_0, y_j)$ can be determined purely geometrically from

$$\rho(h)y_j = e^{2\pi i\Theta_j(h)} y_j, \tag{4.17.b}$$

for any $h \in H_j$ , because $y_j$ is an eigenvector for the linearized flow. Taking linear combinations $y = \sum y_j$ the dynamics, i.e. the eigenvalues $\pm i\beta_j$ themselves, enter via (4.13.$b$) as follows.

Given the symmetries $(H_j, K_j, \Theta_j)$ of the pairs $(x_0, y_j), j \in J$, consider the homomorphism

$$\begin{array}{llll} \underline{\Theta}: & \bigcap_{j\in J} H_j & \longrightarrow & (\mathbb{R}/\mathbb{Z})^{|J|} \\ & h & \mapsto & (\Theta_j(h))_{j\in J} \, . \end{array}$$

Obviously, $ker\ \underline{\Theta} = \bigcap_{j\in J} K_j = \hat{K}$. Moreover

$$\left(\bigcap_{j\in J} H_j\right)/\hat{K} \quad \cong \quad im\ \underline{\Theta}$$

is a compact abelian Lie group. We may view $im\ \underline{\Theta}$ as a lattice in the torus $(\mathbb{R}/\mathbb{Z})^{|J|}$, or in the covering space $\mathbb{R}^{|J|}$. Here we use the term lattice freely, to include the non-discrete cases.

Now let $k_j := \frac{q}{2\pi/\beta_j}$ , $\underline{k} := (k_j)_{j\in J}$ . Then

$$\hat{H} \quad = \quad \underline{\Theta}^{-1}(span\ \underline{k}) \, ,$$

by (4.13.$b$). Put differently,

$$\hat{H}/\hat{K} \quad \cong \quad (span\ \underline{k}) \cap im\ \underline{\Theta} \, .$$

Knowing the eigenvectors $y_j$ , the lattice $im\ \underline{\Theta}$ is fixed independently of the values of the $\beta_j$ . Varying the $\beta_j$ , we can adjust $span\,\underline{k}$ to be any subspace with rationally dependent coefficients. The group $\hat{H}$ varies correspondingly. Fixing $\underline{\Theta}$ and $span\,\underline{k}$, and hence $\hat{H}$, the vector $\underline{k}$ is determined uniquely by $k_j > 0$ and $lcm\,(1/k_j)_{j\in J} = 1$, cf. (4.13.$a$). Finally, this determines $\hat{\Theta}$ by

$$k_j\,\hat{\Theta}(h) \quad \equiv \quad \Theta_j(h),$$

for all $j \in J$, $h \in \hat{H}$.

In this way virtual symmetry is a synthesis between geometry, represented by $\underline{\Theta}$, and dynamics, represented by $\underline{k}$.

**4.10 Remark :**

Let us briefly relate our result on virtual symmetries of a stationary solution $(\lambda_0, x_0)$ to the work of Golubitsky and Stewart [Go&St1, Go&Sch&St]; for more examples in their direction see [Go&St2,3, Mon&Rob&St, Rob&Swi&Wag]. Note that they are interested in sufficient conditions for local equivariant Hopf bifurcation with symmetry $(\hat{H}, \hat{K}, \hat{\Theta})$ whereas lemma 4.8 gives necessary conditions for bifurcation. They consider the case of only one imaginary pair $\pm i$ of eigenvalues ($\bar{j} = j = 1$, $\beta := \beta_j = 1$, $q = 2\pi/\beta$) at $x_0 = 0$, $\Gamma_{x_0} = \Gamma$. The bifurcating branches of periodic solutions then arise from eigenvectors $y$ of $i\beta$ in a fixed point subspace of $\hat{H}^{\hat{\Theta}}$ under the action $\tilde{\rho}$ of $\tilde{\Gamma} = \Gamma \times S^1$ on $X^c$. This action is given by

$$\tilde{\rho}(\gamma, \vartheta)y \quad = \quad e^{-2\pi i\vartheta}\rho(\gamma)y \, .$$

Their crucial condition for bifurcation is that the space of such $\hat{H}^{\hat{\Theta}}$-fixed eigenvectors has complex dimension 1. In our setting this requirement means that

$$\Gamma_{x_0}(\langle y\rangle^c) \quad = \quad \hat{H}$$

is satisfied for a unique complex one-dimensional subspace $\langle y\rangle^c$ of the eigenspace of $i\beta$. For $\hat{K} := \Gamma_{x_0} \cap \Gamma_y$ , $m := |\hat{H}/\hat{K}|$ we obtain from (4.13.c) that

$$\hat{\Theta} \quad \equiv \quad r \qquad\qquad (mod\ m)$$

where $r$ is the representation of $\hat{H}/\hat{K} \cong \mathbb{Z}_m$ on $\langle y\rangle^c$.

By now it is time to remember that virtual periods and virtual symmetries come up in our main theorems 2.9 and 2.10 on global equivariant Hopf bifurcation. Only the next lemma uses analyticity assumption (2.20.c) which enters into these theorems. In (2.20.c) the linearizations $L_l^{K_0}(\lambda)$, restricted to $X^{K_0}$ along the stationary branches $(\lambda, x^l(\lambda)) \in \Lambda \times X^{H_0}$, were required to depend analytically on $\lambda$ near centers. We will obtain a bound $c_0$ on the virtual periods, which is uniform on a dense subset of the set of centers on these branches. This way, we will avoid that our global continua consist only of centers, cf. §7. Denote

$VP(c_0)$ : the set of $(\lambda, x^l(\lambda))$, $0 \le l \le \bar{l}$, $\lambda \in \Lambda$, such that all virtual periods $q$ of $(\lambda, x^l(\lambda))$ for virtual symmetries $(\hat{H}, \hat{K}, \hat{\Theta})$ with $\hat{K} \ge K_0$ satisfy (4.18)

$$q \quad < \quad c_0 \ .$$

**4.11 Lemma :**

*Let nondegeneracy assumption (2.20.a) and analyticity assumption (2.20.c) hold.*

*Then there exists a large $c_0 > 0$ such that the set $VP(c_0)$ is a dense subset of the set of centers*

$$\{\ (\lambda, x^l(\lambda)) \ \mid\ 0 \le l \le \bar{l},\ \lambda \in \Lambda,\ \text{and } L_l^{K_0}(\lambda) \text{ has imaginary eigenvalues }\}\ .$$

**Proof :**

Let us work in $X^{K_0}$. Then the lemma and its proof reduces to [Fie2, lemma 4.8] and the proof given there. □

We conclude this section with a lemma relating the symmetries of a periodic solution $x(t)$ to its virtual symmetries, in case the virtual period equals the minimal period. It turns out that the symmetry of $x(t)$ extends the virtual symmetry, in that case. In some applications we will encounter virtual symmetries which cannot be extended, for group theoretic reasons, cf. §8. In those cases, virtual symmetry coincides with symmetry provided virtual period coincides with minimal period.

**4.12 Lemma :**

*Let $x(t)$ be a periodic solution of (1.1) with minimal period $p$ and symmetry $(H, K, \Theta)$. Let $q = p$ be a virtual period of $x(t)$ with virtual symmetry $(\hat{H}, \hat{K}, \hat{\Theta})$. Then*

$$\hat{H} \leq H \quad \text{and} \quad \hat{\Theta} = \Theta|_{\hat{H}} \quad . \tag{4.19}$$

*In other words*

$$\hat{H}^{\hat{\Theta}} \quad \leq \quad H^{\Theta}; \tag{4.19$'$}$$

*the symmetry extends the virtual symmetry.*

*In particular, $\hat{H} = H$ implies $(\hat{H}, \hat{K}, \hat{\Theta}) = (H, K, \Theta)$.*

**Proof :**

The proof is easy. It only uses definitions 1.1 and 1.2. Let $y(t)$ represent the virtual symmetry $(\hat{H}, \hat{K}, \hat{\Theta})$ and the virtual period $q = p$. Then for all $\hat{h} \in \hat{H}$, $t \in \boldsymbol{R}$

$$(\,\rho(\hat{h})x(t),\ \rho(\hat{h})y(t)\,) \;=\; \hat{h}\,(x(t), y(t)) \;=\; (\,x(t + \hat{\Theta}(\hat{h})p),\ y(t + \hat{\Theta}(\hat{h})p)\,) \quad .$$

Hence $\hat{H} \leq H$. Because $p$ is the minimal period of $x$ and

$$\rho(h)x(t) \quad = \quad x(t + \Theta(h)p) \ , \quad \text{for all } h \in H,\ t \in \boldsymbol{R},$$

we also conclude $\hat{\Theta}(\hat{h}) \equiv \Theta(\hat{h})$ $(mod\ \mathbb{Z})$, i.e. $\hat{\Theta} = \Theta|_{\hat{H}}$. This proves (4.19). The remaining claims follow trivially from (4.19) and the easy proof is complete. □

In case $q \neq p$, lemma 4.12 need not hold. Examples with $q = 2p$ are given by the period doubling bifurcations listed in §5, table 5.2.

# 5 Generic local theory

## 5.1 Outline

In this section we present a reasonably complete list of bifurcations of periodic solutions for equations

$$\dot{x} = f(\lambda, x)$$

with $f$ generic, but equivariant with respect to a cyclic group $G = \mathbb{Z}_n$, $n \leq \infty$. For complicated groups $G$ we are unable to give such a theory because complicated singularities occur which are not sufficiently understood. Fortunately, cyclic equivariance is intimately related to periodic solutions by their symmetry $(H, K, \Theta)$. The "reduction" to cyclic $G$ ($\cong H_0/K_0$) was motivated in §1.5. Our results and proofs are much in the spirit of the Kupka-Smale theorem [Kup, Peix, Smal] and its one-parameter generalizations [Bru, Med, All&M-P&Y2, Fie2], relying heavily on transversality theory.

We proceed as follows. In §5.2 we briefly state a simple result on generic local Hopf bifurcation under $\mathbb{Z}_n$-equivariance, $n \leq \infty$. Three types of secondary bifurcations of rotating and frozen waves are then discussed in §5.3: turn, flop doubling, and freezing. In §5.4, discrete waves provide us with six cases: the turn, three doublings (flip, flop, and flip-flop), and two pitchforks (flip and flip-flop). Fortunately André Vanderbauwhede noticed that two of these cases are essentially equivalent, cf. lemma 5.12. Thus only five cases remain. Concentric waves are a special case of discrete waves which reduces to the case of no symmetry discussed in §3. Our generic results are summarized in §5.5, theorem 5.11 and table 5.2, postponing genericity proofs to §10. In our treatment of discrete waves we favor Poincaré maps, cf. §3.3. For a very elegant treatment of secondary bifurcations in the operator setting (1.23) cf. [Van4]. However, genericity is not tackled there.

## 5.2 Generic centers

The term generic center was introduced in definition 2.2.

**5.1 Lemma :**

*Let $(\lambda_0, x_0)$ be a generic $H$-center with isotropy $G_{x_0} = H \leq \mathbb{Z}_n$ and with representation $r$ of $H$ as explained in §2.2, see (2.4.b). Then the bifurcating local branch $(\lambda(s), x(s,\cdot), p(s))$ of periodic solutions has constant symmetry $(H, K, \Theta)$, near $(\lambda_0, x_0)$, such that for small $|s|$*

$$\Theta \equiv r \qquad (\textit{mod } |H|, \textit{ if } |H| \textit{ is finite}) \tag{5.1}$$

**Proof :**

Of course we could give a direct proof just repeating [Cra&Rab2] in an $H$-equivariant setting. Instead, we do a little exercise on virtual symmetry.

For $|s| > 0$ small enough, the symmetry $(H, K, \Theta)$ of $(\lambda(s), x(s,\cdot))$ is a virtual symmetry of $(\lambda_0, x_0)$ by corollary 4.6. Because $D_x f(\lambda_0, x_0)$ has only one simple pair of purely imaginary eigenvalues $\pm i\beta$ with eigenvectors $y, \bar{y}$, cf. (2.3.a), we conclude that

$$\rho(h)y = e^{2\pi i \Theta(h)} y, \qquad \text{for all } h \in H.$$

Together with

$$\rho(h)y \quad = \quad e^{2\pi i r h} y, \qquad \text{for all } h \in H \tag{2.4.b}$$

this implies

$$\Theta \cdot h \quad \equiv \quad \Theta(h) \quad \equiv \quad r \cdot h \qquad (mod\ \mathbb{Z}), \quad \text{for all } h \in H,$$

where $\Theta$ is represented by an integer as was explained in (1.28). Now (5.1) is immediate and the proof is complete. □

As was promised in §2.2, the term "generic center" is justified by

**5.2 Theorem :**
*Let $G = \mathbb{Z}_n$, $n \leq \infty$. Then all centers are generic centers for $f$ satisfying (2.13), i.e. for generic $G$-equivariant $f$.*

For our notion of genericity see §1.5, definition 1.3 with $H_0 := G$ and $K_0 := \{0\}$. Also $\mathcal{F}$ was defined there, see (1.36).

A proof of theorem 5.2 can be given directly by explicit perturbations or, more elegantly, by transversality theory, see §10.4. A particularly puzzling situation arises for $G = \mathbb{Z}_\infty$ and $H$-centers $(\lambda_0, x_0)$ with $H = G_{x_0}$ discrete. At first sight this case seems to contradict theorem 5.2, because the group $G$ would force $D_x f(\lambda_0, x_0)$ to have an eigenvalue 0. Indeed $x_0$ is on a frozen wave. But generic frozen waves, i.e. freezings, will never be centers as we shall see in the next section, cf. definition 5.3.

## 5.3 Rotating and frozen waves

We begin to discuss secondary bifurcations of periodic solutions $(\lambda_0, x_0(\cdot))$ with minimal period $p$ and symmetry $(H, K, \Theta)$. As before we assume $f \in \mathcal{F}$ is $G$-equivariant, $G = \mathbb{Z}_n$, $n \leq \infty$, cf. assumption (2.2). In this section we specialize to $n = \infty$, to allow for rotating waves ($H = \mathbb{Z}_\infty$, $K$ discrete). In principle, $H$ itself could be discrete though $G = \mathbb{Z}_\infty$ is not. Geometrically, such solutions would provide invariant 2-tori which are foliated by periodic solutions as well as by periodic group orbits. We exclude such solutions here, returning to them in §§8.4.8 and 9.6. The three generic types of secondary bifurcations of rotating and frozen waves, $H = G = \mathbb{Z}_\infty = \mathbb{R}/\mathbb{Z}$ and $K$ discrete, are collected in definitions 5.3 and 5.5 below.

We derive an algebraic equation for rotating waves $x(t)$. For any $h \in H = \mathbb{R}/\mathbb{Z}$ we have

$$\begin{aligned} \rho(h)x(t) \quad = \quad h \cdot x(t) \quad &= \quad x(t + \Theta(h) \cdot p) \\ &= \quad x(t + \Theta \cdot h \cdot p) \end{aligned} \tag{5.2}$$

for some $\Theta \in \mathbb{Z} \setminus \{0\}$. Define infinitesimal rotation by

$$\mathcal{R}x_0 \quad := \quad D_h\, \rho(0) x_0 \ ,$$

cf. (1.19). Then (5.2) implies

$$\mathcal{R}x(t) \quad = \quad \Theta p\, \dot{x}(t) \quad \neq \quad 0.$$

Therefore $x(t)$ is a rotating wave solution if and only if

$$-\alpha \mathcal{R} x_0 \ + \ f(\lambda, x_0) \quad = \quad 0 \tag{5.3}$$

for some $x_0 = x(t_0)$ with $\mathcal{R}x_0 \neq 0$ and $\alpha \in \mathbb{R} \setminus \{0\}$. Note that

$$\alpha = (\Theta p)^{-1}. \tag{5.4}$$

Next suppose $Hx_0$ is a frozen wave, i.e. $f(\lambda, x_0) = 0$ and $G_{x_0}$ is finite. Then $\mathcal{R}x_0 \neq 0$ and $x_0$ satisfies the same equation (5.3) but with $\alpha = 0$. This fits to $p = \infty$ in (5.4), and the usually dreadful complication of period blow-up is resolved into discussing zeros of $\alpha$. The formulation (5.3) was already used by Auchmuty [Au], 1978.

For later reference (cf. §§6 and 10) we briefly calculate the $x$-linearization, along a rotating wave $x(t)$ through $x_0$, of the flow $\psi_t(f, \lambda, x)$ defined by (1.1). Obviously, $y(t) := D_x\psi_t(f, \lambda_0, x_0)y_0$ satisfies

$$\begin{aligned} \dot{y}(t) &= D_x f(\lambda_0, x(t))y(t) \\ y(0) &= y_0 . \end{aligned} \tag{5.5.a}$$

From equivariance of $f$ we easily obtain the solution $y(t)$ and, accordingly,

$$D_x\psi_t(f, \lambda_0, x_0) = \rho(\alpha t)\, exp\Big((-\alpha\mathcal{R} + D_x f(\lambda_0, x_0))t\Big) . \tag{5.5.b}$$

Secondary bifurcations are conveniently described in the setting (5.3) using Poincaré maps or, equivalently, Ljapunov-Schmidt reduction. Fix a Poincaré section $S = \{x_0\} + S'$ in $X = \mathbb{R}^N$, transversely to the group orbit $Hx_0$ of a (rotating or frozen wave) solution $(\alpha_0, \lambda_0, x_0)$ of (5.3), cf. fig. 5.1. It is particularly convenient to choose $S'$ orthogonal to $\mathcal{R}x_0$; this makes $S'$ invariant under the isotropy $K = G_{x_0}$ of $x_0$. Let $P$ denote orthogonal projection onto $S'$; then $Q := 1 - P$ is the orthogonal projection onto the tangent $\mathcal{R}x_0$ of $Hx_0$. Thus the projected equation

$$-\alpha Q\mathcal{R}x + Qf(\lambda, x) = 0 \tag{5.3)$_Q$}$$

can be solved by the implicit function theorem for

$$\alpha = \alpha(f, \lambda, x), \tag{5.6}$$

locally near $(\alpha_0, \lambda_0, x_0)$, because $Q\mathcal{R}x_0 = \mathcal{R}x_0 \neq 0$. To find all rotating or frozen waves, locally, it remains to solve

$$\mathcal{T}(f, \lambda, x) := P[-\alpha(f, \lambda, x)\mathcal{R}x + f(\lambda, x)] = 0, \tag{5.3)$_P$}$$

where

$$\mathcal{T}: \ (\mathcal{F} \times \Lambda \times S)_{loc} \rightarrow S'$$

is equivariant with respect to the action of $K$ on $S'$. Thus secondary bifurcations of rotating and frozen waves are exactly the bifurcations of the zero set of $\mathcal{T}$.

As a first example, we discuss freezing (generic frozen waves), aiming at definition 5.3 below. As we have indicated heuristically in §1.2, at a freezing a rotating wave slows down its speed of rotation and then starts rotating in the opposite direction, reversing the sign of $\Theta$. In view of (5.3), (5.4) this will correspond to a simple zero of $\alpha$ along the solution path of $(5.3)_P$.

Let $Hx_0$ be a frozen wave of $f(\lambda_0, \cdot)$ with symmetry $(H, K, \pm\Theta)$ and Poincaré section $S$ as described above. Assume

the spectrum of $D_x f(\lambda_0, x_0)$ on the imaginary axis consists of only a simple eigenvalue 0. (5.7.a)

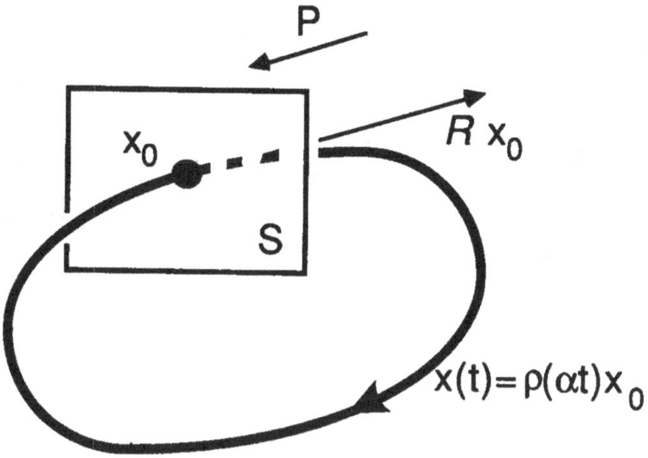


**Fig. 5.1** Poincaré section for a rotating wave

The corresponding eigenvector is $\mathcal{R}x_0$, of course. Equivalently to (5.7.$a$) we may require

$$D_x\mathcal{T}(f,\lambda_0,x_0):\quad S' \quad\to\quad S' \qquad \text{is hyperbolic.} \tag{5.7.$a$)$'$}$$

By (5.7.$a$) the stationary solution $(\lambda_0, x_0)$ is not allowed to be a center. By (5.7.$a$)$'$ solutions of $(5.3)_P$ near $(\lambda_0, x_0)$ lie on a differentiable branch $(\lambda, x_\lambda)$, by the implicit function theorem. We obtain a corresponding map $\lambda \mapsto \alpha(f,\lambda,x_\lambda)$. We require

$$\lambda \quad\mapsto\quad \alpha(f,\lambda,x_\lambda) \qquad \text{has a simple zero at } \lambda=\lambda_0\,. \tag{5.7.$b$}$$

Then the local branch $(\lambda, x_\lambda)$ consists of rotating waves, for $\lambda \neq \lambda_0$, rotating in opposite direction for $\lambda$ on opposite sides of $\lambda_0$. This follows from (5.4) because $\alpha$ changes sign at a simple zero. Moreover, $\Theta$ just changes sign as $\lambda$ passes through $\lambda_0$. Indeed, $\Gamma_{x_\lambda} = K$ is independent of $\lambda$ near $\lambda_0$ by virtual isotropy proposition 4.3 and hyperbolicity assumption (5.7.$a$). Denoting the symmetry of $x_\lambda$ by $(H, K, \Theta_\lambda)$ we conclude that $|\Theta_\lambda| = |K|$ is independent of $\lambda$. Thus $\Theta$ just changes sign at $\lambda_0$, because $\alpha$ does.

**5.3 Definition :**

*Let $Hx_0$ be a frozen wave of $f(\lambda_0,\cdot)$. We call $(\lambda_0, x_0)$ a* **freezing** *if conditions (5.7.a,b) above hold; cf. fig. 5.2.*

Can period doublings of rotating waves conceivably occur? Denote solutions, minimal period, and symmetry on the primary resp. secondary branch by

$$x, p, (H, K, \Theta) \qquad \text{resp.} \qquad \tilde{x}, \tilde{p}, (\tilde{H}, \tilde{K}, \tilde{\Theta})\,.$$

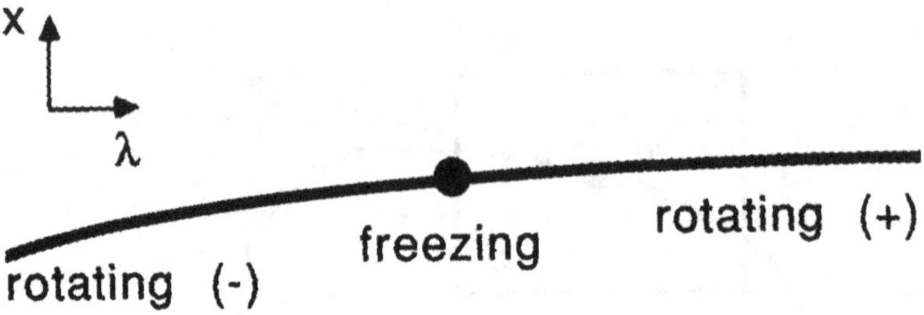


Fig. 5.2 A freezing

Then a period doubling as in definition 3.2 would imply

$$\tilde{H} \;=\; H \;=\; \mathbb{Z}_\infty$$

because we have rotating waves, and

$$\Theta \;=\; 2\tilde{\Theta}$$

because $\tilde{\alpha} = \tilde{\Theta}\tilde{p}$ is close to $\alpha = \Theta p$, $\tilde{p}$ is close to $2p > 0$, and $\Theta, \tilde{\Theta} \in \mathbb{Z}$. Thus $\tilde{K} = ker\tilde{\Theta} = \{h \in \mathbb{R}/\mathbb{Z} \mid \tilde{\Theta} \cdot h \in \mathbb{Z}\} < K$ and

$$[K : \tilde{K}] \;=\; 2\,,$$

i.e. $\tilde{K}$ has index 2 in $K, |K| = |\Theta| = 2|\tilde{\Theta}| = 2|\tilde{K}|$. Thus period doubling of rotating waves can only occur if isotropy breaks from $K$ to $\tilde{K}$. We consider such a symmetry breaking abstractly in definition 5.4 below, calling it a flop. Definition 5.5 will apply this to rotating waves: we obtain a flop doubling.

For the abstract definition of a flop, suppose we are given a $K$-equivariant map $\mathbf{z} \in C^3(\Lambda \times X, X)$ with $K = \mathbb{Z}_k$ and $k$ even. Let $\mathbf{z}(\lambda_0, x_0) = 0$ and suppose $K$ is the isotropy of $x_0$. We assume

the spectrum of $D_x\mathbf{z}(\lambda_0, x_0)$ on the imaginary axis consists of only a simple eigenvalue $\mu(\lambda_0) = 0$. (5.8.a)

Let $y_0$ denote the eigenvector of $\mu(\lambda_0)$, and $\langle y_0 \rangle$ its eigenspace. By equivariance, $K$ leaves $\langle y_0 \rangle$ invariant. Because $K = \{0, 1/k, \cdots, (k-1)/k\}$ has even order there is exactly one nontrivial representation $\rho^-$ of $K$ on $\langle y_0 \rangle$, given by

$$\rho^-(1/k)\, y_0 \;=\; -y_0\,.$$

Assume

$K$ acts on the eigenspace of $\mu(\lambda_0)$ by $\rho^-$. (5.8.b)

Then the restriction

$$D_x\mathbf{z}(\lambda_0, x_0)|_{X^K} \;:\quad X^K \quad\rightarrow\quad X^K$$

is invertible, and we get a primary branch $(\lambda, x_\lambda) \in \Lambda \times X^K$ of zeros of $\mathbf{z}$ for $\lambda$ near $\lambda_0$. Along this branch, the eigenvalue $\mu(\lambda_0)$ continues as a real eigenvalue $\mu(\lambda)$ of $D_x\mathbf{z}(\lambda, x_\lambda)$. We impose the transversality condition

$$D_\lambda \mu(\lambda_0) \quad\neq\quad 0\,. \tag{5.8.c}$$

By the usual local bifurcation theorem [Cra&Rab1], (5.8.$a, c$) imply bifurcation of a local, unique secondary branch

$$(\lambda(s), \tilde{x}(s)) \quad,\quad |s| < \varepsilon\,,$$

of zeros of $\mathbf{z}$. Via (5.8.$b$), our virtual symmetry proposition 4.3 (or directly: equivariant Ljapunov-Schmidt reduction) implies that this branch has isotropy

$$\tilde{K} \quad=\quad ker\,\rho^{\sim} \quad=\quad \{0, 1/\tilde{k}, \cdots, (\tilde{k}-1)/\tilde{k}\}$$

with $\tilde{k} = k/2$. Thus the secondary branch is a pitchfork, i.e. it can be parametrized such that $\lambda(s)$ is even in $s$, and

$$\rho(1/k)\;\tilde{x}(s) \quad=\quad \tilde{x}(-s)\,. \tag{5.9}$$

We finally require a curvature condition

$$D_s^2\lambda(0) \quad\neq\quad 0 \tag{5.8.d}$$

for the secondary branch; cf. fig. 5.3.

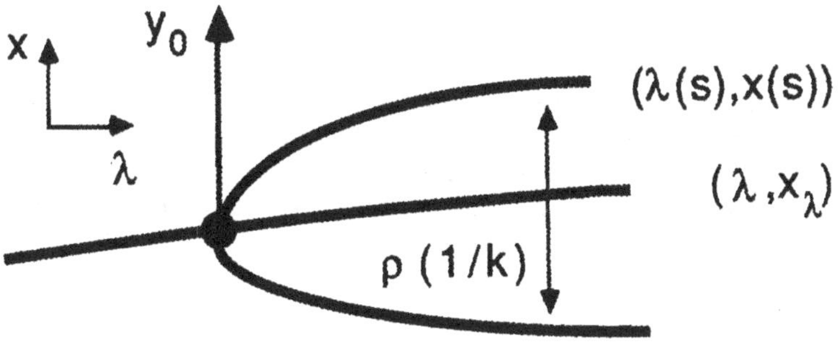


**Fig. 5.3 A flop**

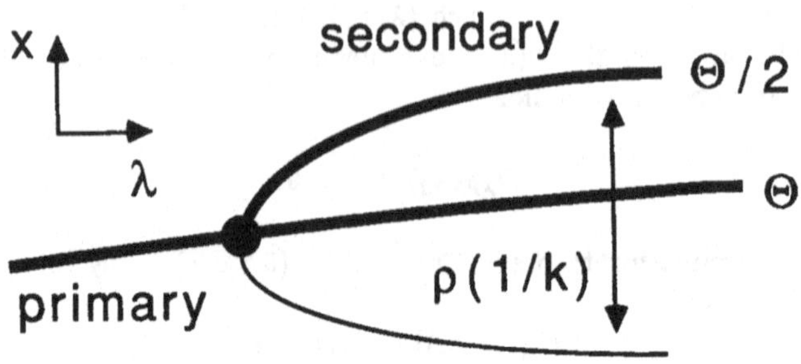


**Fig. 5.4** A flop doubling of rotating waves, $k = |\Theta|$.

**5.4 Definition :**

*Let $(\lambda_0, x_0) \in \Lambda \times X^K$ be a zero of a $K$-equivariant map $\mathbf{z} \in C^3(\Lambda \times X,\ X)$ as above. Then we call the zero $(\lambda_0, x_0)$ a* **flop** *of $\mathbf{z}$ if conditions (5.8.a-d) above hold.*

*For later reference (lemma 5.9 and theorem 5.11(c) below) we also define the following variant here, which is suitable for a fixed point setting. Let $(\lambda_0, x_0) \in \Lambda \times X^K$ be a fixed point of a map $Z \in C^3(\Lambda \times X,\ X)$ which is $K$-equivariant with respect to $x$. Replace condition (5.8.a) by*

$$\text{the spectrum of } D_x Z(\lambda_0, x_0) \text{ on the unit circle consists of only a simple eigenvalue } \mu(\lambda_0) = 1. \tag{5.8.a$'$}$$

*Then the fixed point $(\lambda_0, x_0)$ is called a* **flop** *of $Z$ if conditions (5.8.a)$'$, (5.8.b-d) hold with $\mathbf{z}(\lambda, x) := Z(\lambda, x) - x$.*

Talking about fixed points of $Z$ it is natural to talk about fixed points of iterates of $Z(\lambda, \cdot)$. This accounts for the modified eigenvalue condition (5.8.a)$'$. The reader is encouraged to contrast this definition of a flop with definition 3.2 of a flip. Note that a flop of $Z(\lambda, x)$ constitutes a pitchfork. However, taking $Z(\lambda, x) := \Pi(\lambda, \Pi(\lambda, x))$ for a rotating wave, a flop of $\mathbf{z}(\lambda, x) := \mathcal{T}(f, \lambda, x)$ (cf. $(5.3)_P$) defines a period doubling of $\Pi$, i.e. a "flip" of $\Pi$, if we ignore equivariance with respect to $K$. The word flop, on the other hand, emphasizes this equivariance.

**5.5 Definition :**

**A turn of rotating waves** *is a turn of zeros of $\mathbf{z}(\lambda, x) := \mathcal{T}(f, \lambda, x)$ which are rotating waves, cf. $(5.3)_P$ and definition 3.1. Applying virtual symmetry proposition 4.3 as for the freezing, the symmetry $(H, K, \Theta)$ remains unchanged near a turn of rotating waves.*

**A flop doubling of rotating waves** *is a flop of zeros of $\mathbf{z}(\lambda, x) := \mathcal{T}(f, \lambda, x)$ which are rotating waves with symmetry $(H, K, \Theta)$ on the primary branch and with symmetry $(H, \tilde{K}, \tilde{\Theta})$ on the secondary branch. In particular $K = \mathbb{Z}_{|\Theta|}$ with $\Theta$ even, $\tilde{K} = \mathbb{Z}_{|\Theta|/2}$, and $\tilde{\Theta} = \Theta/2$; cf. fig. 5.4 and definition 5.4.*

## 5.4 Concentric and discrete waves

Secondary bifurcations of discrete waves provide the largest number of distinct cases. We consider $f \in \mathcal{F}$, $G$-equivariant, $G = \mathbb{Z}_n$, cf. assumption (2.2). Except for the next definition where $n = \infty$, we assume in this section that $n$ is finite.

Consider concentric waves with $H = G = \mathbb{Z}_\infty$ first. Such waves have symmetry $(H, K, \Theta) = (\mathbb{Z}_\infty, \mathbb{Z}_\infty, 0)$. They lie entirely in $X^G$, where $G$ acts trivially.

**5.6 Definition :**

*We call a concentric wave* $(\lambda_0, x_0)$ *in* $X^G, G = \mathbb{Z}_\infty$, *a* **turn** *resp. a* **flip doubling** *if it is a turn resp. a flip doubling in the sense of definition 3.2 for the flow restricted to* $X^G$. *But in addition, we require that the Floquet multipliers of* $(\lambda_0, x_0)$ *on the unit circle, with respect to* $X$, *equal those of the restriction of the flow to* $X^G$, *with equal multiplicities.*

The additional assumption guarantees that any bifurcating periodic solutions remain in $X^G$, remaining concentric waves.

Now we consider finite $G$ and discrete waves. For a geometric classification, as well as for genericity proofs we introduce the following notation, suppressing the parameter $\lambda$ for a moment. For a periodic solution $x_0(t)$ with minimal period $p$ and symmetry $(H, K, \Theta)$, $G = \mathbb{Z}_n$ , $H = \{0, \frac{1}{\mathbf{h}}, \cdots, \frac{\mathbf{h}-1}{\mathbf{h}}\}$, $K = \{0, \frac{1}{k}, \cdots, \frac{k-1}{k}\}$, $\Theta \in \mathbb{Z}\,(mod\ \mathbf{h})$ we trivially observe

$$\frac{1}{k} = \frac{m}{\mathbf{h}} \quad \text{for} \quad m := |H/K|\,. \tag{5.10}$$

As before, we choose a Poincaré section $S = \{x_0\} + S'$ to $x(t)$, say at $x_0 := x(0)$, such that $K$ leaves $S$ invariant. This certainly holds for $S' = \langle \dot{x}_0 \rangle^\perp$. Next we choose $h \in H$ such that

$$h \text{ generates } H, \text{and} \tag{5.11.a}$$

$$\Theta(h) = 1/m\,. \tag{5.11.b}$$

In particular $\Theta(h)$ generates $im\ \Theta$, and

$$h_0 := mh \quad \text{generates} \quad K = ker\,\Theta\,. \tag{5.11.c}$$

Note here that $G$ is written additively, while the representation $\rho$ is written multiplicatively; e.g. $\rho(mh) = \rho(h)^m$. Later on it will become important that $h$ is not at all determined uniquely by conditions (5.11.$a, b$), see lemma 5.12 below. For the moment, however, we consider $h$ as given and fixed. Applying $h$ to our Poincaré section S we can define further Poincaré sections $S_j$ to the same solution $x_0(t)$ by

$$S_j := \rho(h)^j\, S\,, \qquad 0 \le j \le m\,.$$

Obviously $S_0 = S_m = S$, because $K$ leaves $S$ invariant. These sections $S_j$ yield "Poincaré maps"

$$\hat{\Pi}_j : (S_j)_{loc} \to S_{j+1}\,,$$

mapping $x \in S_j$ to the point where its trajectory first hits $S_{j+1}$ in positive time, as usual (cf. fig. 5.5). This occurs at the "Poincaré time" $\tau_j = \tau_j(x)$. Let

$$\Pi : S_{loc} \to S$$

denote the usual Poincaré-map and put

$$\Pi_0 := \rho(h)^{-1} \hat{\Pi}_0 : S_{loc} \to S \tag{5.12}$$

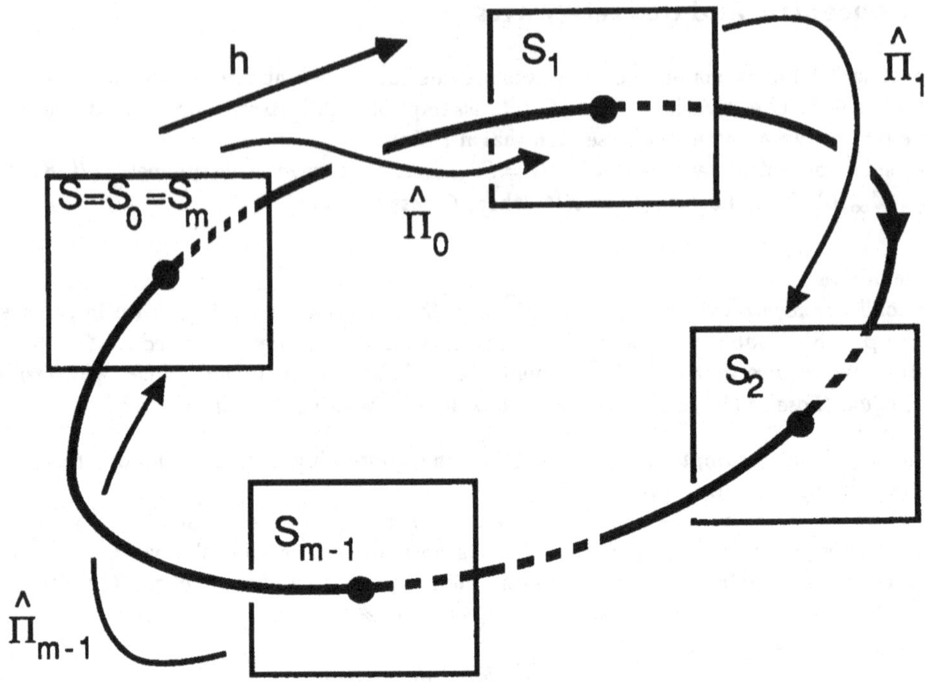


**Fig. 5.5** Poincaré-maps for discrete symmetry

**5.7 Lemma :**
*With the above notation we obtain*

$$\Pi \quad = \quad \rho(h_0)\Pi_0^m ,$$

*where* $\Pi_0^m = \Pi_0 \circ \cdots \circ \Pi_0$ *denotes the m-th iterate.*

**Proof :**
From $G$-equivariance of the flow (1.1) we get

$$\hat{\Pi}_j \quad = \quad \rho(h)^j \hat{\Pi}_0 \rho(h)^{-j} \quad = \quad \rho(h)^{j+1} \Pi_0 \rho(h)^{-j} .$$

On the other hand

$$\begin{aligned} \Pi \quad &= \quad \hat{\Pi}_{m-1} \circ \hat{\Pi}_{m-2} \circ \cdots \circ \hat{\Pi}_1 \circ \hat{\Pi}_0 \quad = \\ &= \quad \rho(h)^m \Pi_0 \rho(h)^{-(m-1)} \; \rho(h)^{(m-1)} \Pi_0 \rho(h)^{-(m-2)} \circ \cdots \circ \rho(h)^2 \Pi_0 \rho(h)^{-1} \; \rho(h) \Pi_0 \quad = \\ &= \quad \rho(h)^m \Pi_0^m \quad = \quad \rho(h_0)\Pi_0^m , \end{aligned}$$

completing the proof. □

Note that $\Pi_0$ is $K$-equivariant. We will prove in §10 that the only generic bifurcation of $\Pi_0$ are turns, flips, flops, and flip-flops. See definition 5.8 and lemma 5.9 below. Via $\Pi = \rho(h_0)\Pi_0^m$ we then obtain a complete list of bifurcations for the Poincaré map $\Pi$ (and its iterates $\Pi^i$) because the sets "*Fix*" of fixed points satisfy

$$\bigcup_{i \in \mathbb{N}} Fix\ \Pi^i \quad = \quad \bigcup_{i' \in \mathbb{N}} Fix\ \Pi_0^{i'} .$$

We work out the details and summarize our claims in §5.5, see theorem 5.11 and table 5.2.

To define a flip-flop bifurcation let $Z \in C^3(\Lambda \times X,\, X)$, $K$-equivariant with respect to $x$, have a fixed point $(\lambda_0, x_0) \in \Lambda \times X^K$:

$$Z(\lambda_0, x_0) \quad = \quad x_0 .$$

We think of $Z$ as a "Poincaré" map $\Pi_0$, but give an abstract definition. To account for the "flip" part we assume

$$\text{the spectrum of } D_x Z(\lambda_0, x_0) \text{ on the unit circle consists of only a simple eigenvalue } \mu(\lambda_0) = -1. \tag{5.13.a}$$

Thus we may hope for fixed points of $Z(\lambda, Z(\lambda, \cdot))$ to bifurcate at $(\lambda_0, x_0)$ from the local primary branch $(\lambda, x_\lambda) \in \Lambda \times X^K$ of $Z$-fixed points. To account for the "flop" part, we assume that

$$K \text{ acts by } \rho^- \text{ on } \langle y_0 \rangle , \tag{5.13.b}$$

where $K = \mathbb{Z}_k$ with even $k$ and $\rho^-$ is the nontrivial scalar representation of $K$ as in (5.8.$b$). For the continuation $\mu(\lambda)$ of $\mu(\lambda_0)$ along $(\lambda, x_\lambda)$ we again impose the transversality condition

$$D_\lambda \mu(\lambda_0) \quad \neq \quad 0 . \tag{5.13.c}$$

As usual, [Cra&Rab1] then implies a pitchfork at $(\lambda_0, x_0)$ with a bifurcating local, unique branch

$$(\lambda(s),\ x(s)), \qquad\qquad |s| < \varepsilon ,$$

of fixed points of $Z(\lambda, Z(\lambda, \cdot))$. Again, we may parametrize the secondary branch over $\langle y_0 \rangle$ such that

$$\begin{aligned} \rho(1/k)x(s) \quad &= \quad x(-s) , \\ Z(\lambda(s), x(s)) \quad &= \quad x(-s) . \end{aligned}$$

In addition, we require a curvature condition

$$D_s^2 \lambda(0) \quad \neq \quad 0 \tag{5.13.d}$$

for this secondary branch; cf. fig. 5.6.

**5.8 Definition :**

*Let $(\lambda_0, x_0) \in \Lambda \times X^K$ be a fixed point of a $K$-equivariant map $Z \in C^3(\Lambda \times X,\, X)$. We call $(\lambda_0, x_0)$* **type 0, turn,** *or* **flip,** *if it is type 0, turn, or flip, respectively, for the restriction of $Z$ to $\Lambda \times X^K$, cf. definition 3.2. But in addition, we require that the spectrum of $D_x Z(\lambda_0, x_0)$ on the unit circle equals that spectrum of the restriction of $D_x Z(\lambda_0, x_0)$ to $X^K$, with equal multiplicities . In particular, this excludes symmetry breaking bifurcations of fixed points of $Z$ and its iterates.*

*We call $(\lambda_0, x_0)$ a* **flip-flop,** *if conditions (5.13.a-d) above hold.*

Utilizing also definition 5.4 of a flop of $Z := \Pi_0$, we can now express genericity for the "Poincaré" map $\Pi_0$, defined in (5.12).

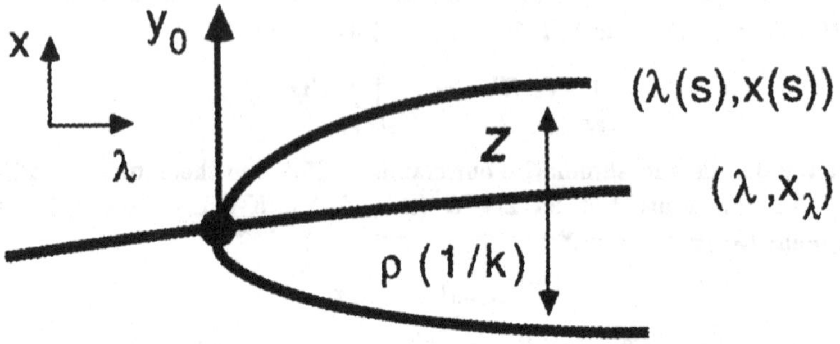


**Fig. 5.6** A flip-flop.

**5.9 Lemma :**

*Let $(\lambda_0, x_0)$ be on a periodic solution $(\lambda_0, x(t))$ of $f$. Then, for generic $f$, one of the following holds for the fixed point $(\lambda_0, x_0)$ of $Z = \Pi_0$ :*

| | |
|---|---|
| *Case c.0* : | $(\lambda_0, x_0)$ *is of type 0* |
| *Case c.1.1* : | $(\lambda_0, x_0)$ *is a turn* |
| *Case c.1.2* : | $(\lambda_0, x_0)$ *is a flop* |
| *Case c.2.1* : | $(\lambda_0, x_0)$ *is a flip* |
| *Case c.2.2* : | $(\lambda_0, x_0)$ *is a flip-flop,* |

*where $K := G_{x_0}$ in definitions 5.4 and 5.8 above.*

The proof of this lemma requires transversality theory and is deferred to §10.

Let us return now to the original (full) Poincaré map $\Pi$ which relates to $\Pi_0$ by

$$\Pi = \rho(h_0)\Pi_0^m$$

(lemma 5.7), where $h_0 = mh$ generates $K$ as in (5.11.a-c). Our classification of bifurcations of $\Pi_0$ leads to a corresponding classification of bifurcations of $\Pi$.

**5.10 Definition :**

*We introduce a terminology for periodic solutions $(\lambda_0, x_0)$ of (1.1) with symmetry $(H, K, \Theta)$, viewing $(\lambda_0, x_0)$ as a fixed point of its Poincaré map $\Pi$. Given the type of $(\lambda_0, x_0)$ as a fixed point of $\Pi_0$, and given $m = |H/K|$, a name is assigned to $(\lambda_0, x_0)$ according to table 5.1. These names are justified in theorem 5.11 and in table 5.2 below.*

| Case | $\Pi_0$ | $\Pi$ |
|---|---|---|
| c.0 | type 0 | type 0 |
| c.1.1 | turn | turn |
| c.1.2 | flop | flop doubling |
| c.2.1 $m$ odd | flip | flip doubling |
| c.2.1 $m$ even | flip | flip pitchfork |
| c.2.2 $m$ odd | flip-flop | flip-flop pitchfork |
| c.2.2 $m$ even | flip-flop | flip-flop doubling |

*Table 5.1: Terminology of generic secondary bifurcations of discrete waves.*

## 5.5 Generic secondary bifurcations

After the admittedly lengthy explanations and definitions of rotating, frozen, concentric, and discrete waves we can now state the main result of our generic local theory for cyclic groups $G = \mathbb{Z}_n$, $n \leq \infty$. The main result, theorem 5.11, embraces many cases, which are all listed in table 5.2.

Let us explain table 5.2. We consider a periodic solution $(\lambda_0, x_0)$ with symmetry $(H, K, \Theta)$. Various possibilities of generic bifurcations arise. Case (a) summarizes those for which $K = H = G = \mathbb{Z}_\infty$. Case (b) considers rotating and frozen waves, i.e. $K \neq H = G = \mathbb{Z}_\infty$. And case (c) considers $H \leq G = \mathbb{Z}_n$, $n < \infty$. We skip the possibility that $G = \mathbb{Z}_\infty$ while $H$ is finite, i.e. $|G/H| = \infty$. This one remaining case would lead to invariant tori which are foliated periodically by $G$-orbits. We postpone a discussion to §§8.3.8 and 9.5.

Each of the cases (a-c) splits into several subcases, and we specify the names of the respective bifurcations indicating the numbers of the appropriate definitions and figures. The remaining columns denote the jumps in minimal period and symmetry from the primary branch to the secondary branch with minimal period $\tilde{p}$ and symmetry $(\tilde{H}, \tilde{K}, \tilde{\Theta})$. The quotient $\tilde{p}/p$ is understood as the limit at the bifurcation point, of course. Because $\tilde{H}, \tilde{K}$ are subgroups of $H, K$ we just give their index $[H : \tilde{H}]$ resp. $[K : \tilde{K}]$, to indicate symmetry breaking. Redundantly we include $m/\tilde{m} = |H/K|/|\tilde{H}/\tilde{K}| = [H : \tilde{H}]/[K : \tilde{K}]$, when $m$ and $\tilde{m}$ are finite. The final two columns indicate the subtleties of symmetry breaking in case $G = \mathbb{Z}_n$ is finite. As in (5.11.*a*-c) we choose $\tilde{h}$, generating $\tilde{H}$, such that $\tilde{\Theta}(\tilde{h}) = 1/\tilde{m}$, analogously to $h$. We relate $\Theta$ to $\tilde{\Theta}$ in two ways. First we give $\tilde{\Theta}(h)$ or, if $[H : \tilde{H}] = 2$, $\tilde{\Theta}(2h)$. This defines the phase shift induced by the generator $h$ or $2h$ on the secondary branch. Then we supply an $\tilde{h}$, in terms of $h$, which induces a minimal phase shift of $1/\tilde{m}$ on the secondary branch. To do this, we decompose $|K| = k$ as

$$k \quad = \quad 2^{\kappa} \cdot k' \quad \text{with } k' \text{ odd.}$$

**5.11 Theorem :**

*Let assumption (2.13) hold for $f \in \mathcal{F}$, i.e. $f$ is generic $G$-equivariant, $G = \mathbb{Z}_n$, $n \leq \infty$. Let $(\lambda_0, x_0)$ be any periodic solution of (1.1) with symmetry $(H, K, \Theta)$. Assume $G/H$ is finite.*

*Then either $(\lambda_0, x_0)$ is of type 0 (cf. definition 3.2), or $(\lambda_0, x_0)$ occurs among the bifurcation types listed in table 5.2.*

| wave type | case | m | bifurcation | definitions | fig. | $\tilde{p}/p$ | $[H:\tilde{H}]$ | $[K:\tilde{K}]$ | $m/\tilde{m}$ | $\tilde{\Theta}$ | $\tilde{h}$ |
|---|---|---|---|---|---|---|---|---|---|---|---|
| $K=H=G=Z_\infty$ | a.1 | 1 | turn | 5.6,3.2,3.1 | 3.1 | 1 | 1 | 1 | 1 | $\tilde{\Theta}=\Theta=0$ | - |
| concentric | a.2 | 1 | flip doubling | 5.6,3.2 | 3.2 | 2 | 1 | 1 | 1 | $\tilde{\Theta}=\Theta=0$ | - |
| $K\neq H=G=Z_\infty$ | b.1 | $\infty$ | turn | 5.5,3.1 | 3.1 | 1 | 1 | 1 | - | $\Theta$ | - |
| rotating & | b.2 | $\infty$ | flop doubling | 5.5,5.4 | 5.4 | 2 | 1 | 2 | - | $\Theta/2$ | - |
| frozen | b.3 | $\infty$ | freezing | 5.3 | 5.2 | - | 1 | 1 | - | $-\Theta$ | - |
| $G=Z_n$, $n<\infty$ | c.1.1 | any | turn | 5.10,5.8 3.2, 3.1 | 5.7 | 1 | 1 | 1 | 1 | $\Theta$ | h |
| concentric | c.1.2 | any | flop doubling | 5.10,5.4 | 5.7 | 2 | 1 | 2 | 1/2 | $\tilde{\Theta}(h)=1/2m$ | h |
| | c.2.1 | odd | flip doubling | 5.10,5.8 3.2 | 5.7 | 2 | 1 | 1 | 1 | $\tilde{\Theta}(h)=1/2 + 1/2m$ | $(k'm+2)h$ |
| & | | even | flip pitchfork | 5.10,5.8 3.2 | 5.7 | 1 | 2 | 1 | 2 | $\tilde{\Theta}(2h)=2/m$ | 2h |
| discrete | c.2.2 | odd | flip-flop pitchfork | 5.10,5.8 | 5.7 | 1 | 2 | 2 | 1 | $\tilde{\Theta}(2h)=2/m$ | $(3k'm+1)h$ if $4 \mid (k'm+1)$; $(k'm+1)h$ otherwise |
| | | even | flip-flop doubling | 5.10,5.8 | 5.7 | 2 | 1 | 2 | 1/2 | $\tilde{\Theta}(h)=1/2 + 1/2m$ | $(k'm+1)h$ |

**Table 5.2: Generic secondary bifurcations**

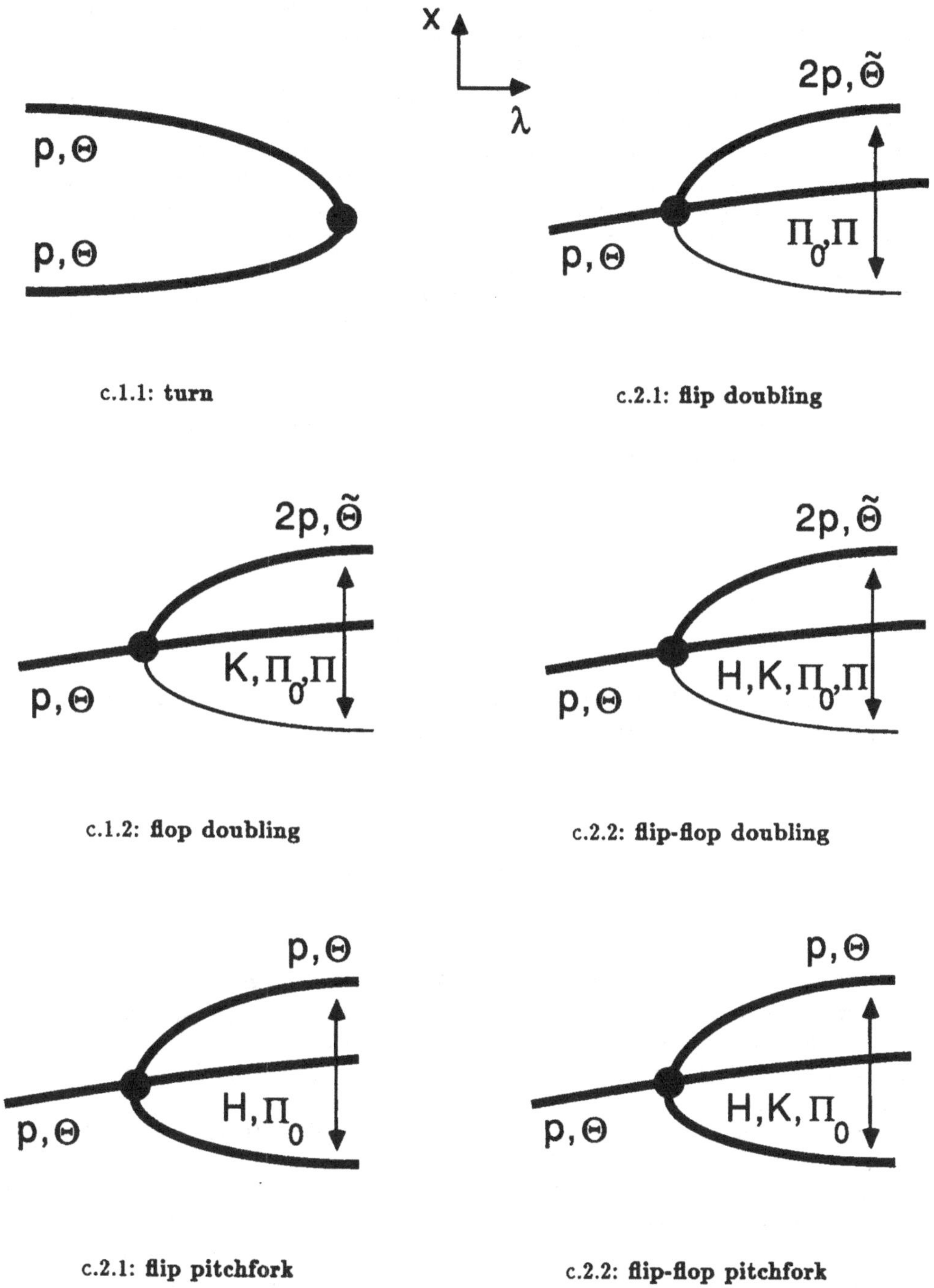


**Fig. 5.7** Geometry of generic secondary bifurcations for $G = \mathbb{Z}_n$, $n < \infty$, cf. theorem 5.11.

It is not particularly difficult to derive normal forms for turns, flops, flips and flip-flops of $\Pi_0$, using e.g. the results in [Van1], [Go&Sch]. After a glance at the perturbation results for $\Pi_0$ in §10, this provides examples for each of the less obvious bifurcations listed in table 5.2 (c). They all occur, and they are all claimed to be generic. For nonlinear $\mathbb{Z}_2$-action ($n = 2$) the cases of a turn and of a flip pitchfork were discussed by [Klič], already.

To prove theorem 5.11 we have to establish that table 5.2 gives a complete list of generic bifurcations of periodic solutions. Again, this is deferred to §10 so that our line of reasoning is not interrupted by a pile of technicalities. For now we only consider case (c), taking the generic bifurcations of $\Pi_0$ for granted as given in lemma 5.9. The reason for this is two-fold: from a transversality point of view lemma 5.9 is the thing to prove, rather than case (c) of table 5.2. And deriving case (c) from lemma 5.9 will be a nice warm-up for §6.

**Proof of theorem 5.11, case (c) :**

Given the bifurcation for $\Pi_0$ at $(\lambda_0, x_0)$ and the action of $h_0 = mh$ on the secondary branch, the bifurcation for $\Pi$ is completely determined because $\Pi = \rho(h_0)\Pi_0^m$ by lemma 5.7. Applying lemma 5.9, it remains to verify the information contained in table 5.2(c) for each of the cases (c.1.1-2.2) of lemma 5.9 and definition 5.10.

**Case c.1.1, $(\lambda_0, x_0)$ is a turn :**

Then we have a turn of $\Pi_0$ within $X^K$. Thus

$$\Pi(\lambda, x) \quad = \quad \rho(h_0)[\Pi_0(\lambda, \cdot)]^m(x) \quad = \quad x$$

for the $\Pi_0$-fixed points $(\lambda, x)$ of the turn of $\Pi_0$ through $(\lambda_0, x_0)$. These $(\lambda, x)$ constitute all fixed points of any iterate of $\Pi$. Thus $H, K, \Theta, m$ remain unchanged.

**Case c.1.2, $(\lambda_0, x_0)$ is a flop :**

For the secondary branch $(\lambda(s), x(s))$ we then have

$$\begin{aligned} \Pi(\lambda(s), x(s)) \quad &= \quad \rho(h_0)[\Pi_0(\lambda(s), \cdot)]^m(x(s)) \quad = \quad \rho(h_0)x(s) \quad = \\ &= \quad x(-s) \quad \neq \quad x(s)\,, \qquad\qquad \text{for } s \neq 0, \end{aligned}$$

because $h_0$ generates $K$ and $K$ acts by $\rho^-$, see (5.8.$b$), (5.9). Thus $\Pi$ sees a period doubling.

Next we claim that $\tilde{H} = H$. Indeed, $x(s) = \Pi_0(\lambda(s), x(s)) = h^{-1}\hat{\Pi}_0(\lambda(s), x(s))$, where $\hat{\Pi}_0$ is the Poincaré map from $S_0$ to $S_1$ with associated Poincaré time $\tau_0$, cf. (5.12). Consequently, $hx(s) = \hat{\Pi}_0(\lambda(s), x(s))$ and hence $h \in \tilde{H}$. Because $h$ generates $H$ and because $\tilde{H} \leq H$ we conclude $\tilde{H} = H$.

We claim that $\tilde{K} = \langle 2h_0 \rangle$ has index 2 in $K = \langle h_0 \rangle$. Indeed $\rho(h_0)x(s) = x(-s) \neq x(s)$ whereas $\rho(2h_0)x(s) = \rho(h_0)^2x(s) = x(s)$. This proves the claim.

It remains to show that $\tilde{\Theta}(h) = 1/2m$. Denoting the minimal period of $(\lambda(s), x(s))$ by $\tilde{p}(s)$ we indeed conclude from the above

$$\tilde{\Theta}(h) \quad = \quad \tau_0(\lambda(s), x(s))/\tilde{p}(s) \quad = \quad (\tilde{p}(s)/2m)/\tilde{p}(s) \quad = \quad 1/2m\,.$$

Because $m/\tilde{m} = [H : \tilde{H}]/[K : \tilde{K}] = 1/2$, we may thus take $\tilde{h} := h$ to obtain $\tilde{\Theta}(\tilde{h}) = 1/\tilde{m}$. This proves all our claims in case c.1.2.

**Case c.2.1, $(\lambda_0, x_0)$ is a flip, $m$ is odd :**
Because $\Pi_0$ is a flip in $X^K$, the secondary branch satisfies

$$\begin{aligned} \Pi_0(\lambda(s), x(s)) &= x(-s) \neq x(s), \quad \text{for } s \neq 0, \text{ and} \\ \rho(h_0)x(s) &= x(s). \end{aligned}$$

Because $m$ is odd, this implies

$$\begin{aligned} \Pi(\lambda(s), x(s)) &= \rho(h_0)[\Pi_0(\lambda(s), \cdot)]^m(x(s)) = \\ &= \rho(h_0)x(-s) = x(-s) \neq x(s), \end{aligned}$$

and $\Pi$ sees a period-doubling. Also

$$\rho(h)x(s) = \hat{\Pi}_0(\lambda(s), x(-s)) = \hat{\Pi}_0(\lambda(s), \cdot) \circ \Pi(\lambda(s), \cdot)\,(x(s))$$

lies on the same time-orbit as $x(s)$, and we get $\tilde{H} = H$. Obviously $\tilde{K} = K$. As before, the Poincaré time $\tau_0$ of $(\lambda(s), x(s))$ associated to $\hat{\Pi}_0$ is given by

$$\tau_0 = \tilde{p}(s)/2m\,,$$

and $\Pi$ takes a Poincaré time of

$$\tilde{p}(s)/2.$$

For the periodic solution $x(s,t)$ through $x(s) = x(s,0)$ this implies

$$\begin{aligned} \rho(h)x(s,0) &= \hat{\Pi}_0(\lambda(s), \Pi(\lambda(s), x(s))) = \\ &= x\left(s, \frac{\tilde{p}(s)}{2} + \frac{\tilde{p}(s)}{2m}\right) \end{aligned}$$

and we conclude that

$$\tilde{\Theta}(h) = \frac{1}{2m} + \frac{1}{2}.$$

It remains to verify that $\tilde{h} := (k'm+2)h$ satisfies conditions (5.11.$a,b$) for $\tilde{H} = H$ and $\tilde{m} = |\tilde{H}/\tilde{K}| = |H/K| = m$ replacing $H$ and $m$. Indeed, $k'm+2$ and $|\tilde{H}| = |H| = |K|\cdot|H/K| = 2^\kappa k'm$ are relatively prime:

$$\begin{aligned} (k'm+2, |\tilde{H}|) &= (k'm+2, 2^\kappa k'm) = \\ &= (k'm+2, k'm) = \\ &= (2, k'm) = 1, \end{aligned}$$

all because $k'm$ is odd. Here $(\cdot,\cdot)$ denotes the greatest common divisor. Thus there exists $j$ such that $j(k'm+2) \equiv 1 \ (mod\ |H|)$, and consequently $\tilde{h}$ generates $\tilde{H} = H$:

$$h = j\tilde{h} \in \langle \tilde{h} \rangle\,.$$

This verifies (5.11.$a$). To verify (5.11.$b$) we calculate $(mod\ \mathbb{Z})$

$$\begin{aligned} \tilde{\Theta}(\tilde{h}) &= (k'm+2)\tilde{\Theta}(h) = (k'm+2)\left(\frac{1}{2m} + \frac{1}{2}\right) \\ &= k'\left(\frac{1}{2} + \frac{m}{2}\right) + \frac{1}{m} + 1 \equiv \frac{1}{m}, \end{aligned}$$

using again that $m$ is odd.

**Case c.2.2, $(\lambda_0, x_0)$ is a flip-flop, $m$ is even :**
Reversing only the sign

$$\rho(h_0)x(s) \quad = \quad x(-s) \quad \neq \quad x(s), \qquad \text{for } s \neq 0,$$

this case is so analogous to the previous one, that we cannot resist the temptation to skip all details.

**Case c.2.2, $(\lambda_0, x_0)$ is a flip-flop, $m$ is odd :**
Because $\Pi_0$ is a flip-flop, the secondary branch satisfies

$$\begin{aligned} \Pi_0(\lambda(s), x(s)) &= x(-s) \neq x(s) \qquad \text{and} \\ \rho(h_0)x(s) &= x(-s) \neq x(s), \qquad \text{for } s \neq 0. \end{aligned}$$

This implies

$$\begin{aligned} \Pi(\lambda(s), x(s)) &= \rho(h_0)[\Pi_0(\lambda(s), \cdot)]^m(x(s)) = \\ &= x((-1)^{m+1}s) = x(s), \end{aligned}$$

and $\Pi$ sees a pitchfork of fixed points. Therefore

$$\rho(h)x(s) \quad = \quad \hat{\Pi}_0(\hat{\lambda}(s), x(-s))$$

does not lie on the periodic solution $x(s,t)$ through $x(s,0) = x(s)$. But

$$\begin{aligned} \rho(h)^2 x(s) &= \rho(h)^2[\Pi_0(\lambda(s), \cdot)]^2(x(s)) = \\ &= \hat{\Pi}_1\left(\lambda(s), \hat{\Pi}_0(\lambda(s), x(s))\right) \end{aligned}$$

does lie on the periodic solution $x(s,t)$. Hence $\tilde{H}$ is generated by $2h$ and $[H : \tilde{H}] = 2$. Obviously $[K : \tilde{K}] = 2$ and $\tilde{m} = m$. This time, the Poincaré time $\tau_0 + \tau_1$ of $(\lambda(s), x(s))$ associated to $\hat{\Pi}_1(\lambda(s), \cdot) \circ \hat{\Pi}_0(\lambda(s), \cdot)$ is given by

$$\tau_0 + \tau_1 \quad = \quad \tilde{p}(s)/(m/2)$$

and $\Pi$ takes a Poincaré time of $\tilde{p}(s)$. For $x(s,t)$ this implies

$$\begin{aligned} \rho(h)^2 x(s,0) &= \hat{\Pi}_1\left(\lambda(s), \hat{\Pi}_0(\lambda(s), x(s))\right) = \\ &= x(s, \tau_0 + \tau_1) = x(s, \frac{2}{m}\tilde{p}(s)) \end{aligned}$$

and we conclude

$$\tilde{\Theta}(2h) \quad = \quad \frac{2}{m}.$$

It remains to verify our claim on $\tilde{h}$. First suppose $(k'm+1)/2$ is odd. Then

$$\begin{aligned} (k'm+1, |H|) &= (k'm+1, 2^\kappa k'm) = \\ &= (k'm+1, 2^\kappa) = 2, \end{aligned}$$

because $\kappa \geq 1$ $(|K| = k = 2^\kappa k'$ is even, cf(5.8.$b$)). If on the other hand $(k'm+1)/2$ is even, then $(3k'm+1)/2$ is odd. As before we obtain

$$\begin{aligned}(3k'm+1, |H|) &= (3k'm+1, 2^\kappa k'm) &= \\ &= (3k'm+1, 2^\kappa) &= 2\,.\end{aligned}$$

Thus, there exists $j$ such that $j(k'm+1) \equiv 2 \ (mod\ |H|)$ resp. $j(3k'm+1) \equiv 2\ (mod|H|)$. In any case, $\tilde{h}$ generates $\tilde{H} = \langle 2h \rangle$ because

$$2h \quad = \quad j\tilde{h} \quad \in \quad \langle \tilde{h} \rangle,$$

which verifies (5.11.$a$). To verify (5.11.$b$) we calculate $(mod\ \mathbb{Z})$ for $\tilde{h} := (\iota m + 1)h$ with appropriate $\iota \in \{1,3\}$ as above

$$\begin{aligned}\tilde{\Theta}(\tilde{h}) &= \tfrac{\iota k'm+1}{2}\,\tilde{\Theta}(2h) &= (\iota k'm+1)/m &= \\ &\equiv 1/m &= 1/\tilde{m}\,.\end{aligned}$$

**Case c.2.1, $(\lambda_0, x_0)$ is a flip, $m$ is even :**
This case is analogous to, but simpler than, the previous one. The simplification arises because $H$ breaks but $K$ does not, so that $\tilde{m} = m/2$ and we may simply take $\tilde{h} = 2h$. Omitting all further details, the proof is complete. □

Following a hint by André Vanderbauwhede [Van4] we should now wonder whether any generic bifurcation orbit $(\lambda_0, x_0)$ corresponds to exactly one case from table 5.2, or whether different cases may in fact be equivalent geometrically. Obviously, the geometries of turns, pitchforks and doublings, as well as the geometries of concentric $(H = \mathbb{Z}_\infty)$, rotating, frozen and discrete waves are mutually distinct. Investigating secondary symmetries $\tilde{H}, \tilde{K}$ of discrete waves we discover that the only possible candidates for equivalent bifurcation orbits are the flop doubling and the flip-flop doubling, when $m$ is even.

**5.12 Lemma :**
*Let $f$ be generic as in theorem 5.11, case (c) and let $m$ be even. Then flop doubling and flip-flop doubling of $(\lambda_0, x_0)$ are geometrically equivalent. More precisely, let $h$ satisfy (5.11.a-c) and suppose $(\lambda_0, x_0)$ is a flop doubling. Then there exists $h'$ satisfying (5.11.a-c) such that $(\lambda_0, x_0)$ becomes a flip-flop doubling if we replace $h$ by $h'$. And vice versa.*

**Proof :**
First suppose that $(\lambda_0, x_0)$ is a flop doubling. Let $h \in H$ generate $H = \tilde{H}$ with $\Theta(h) = 1/m$ and $\tilde{\Theta}(h) = 1/2m$. It is sufficient to construct $h'$, generating $H$, such that

$$\Theta(h') = \frac{1}{m}, \quad \text{and} \tag{5.14.a}$$

$$\tilde{\Theta}(h') = \frac{1}{2m} + \frac{1}{2}\,. \tag{5.14.b}$$

To achieve this, we first write $|K| = 2^\kappa \cdot k'$ with $k'$ odd, $|H| = 2^\kappa k'm$, and observe that

$$\begin{aligned}(k'm+1, |H|) &= (k'm+1, 2^\kappa k'm) &= \\ &= (k'm+1, 2^\kappa) &= 1,\end{aligned}$$

where $(\cdot,\cdot)$ denotes the greatest common divisor, as before. Thus there exists an integer $j \in \mathbb{N}$ such that

$$j \cdot (k'm+1) \quad \equiv \quad 1 \qquad (mod\ |H|).$$

Define

$$h' \quad := \quad (k'm+1)h.$$

Then $h'$ generates $H = \langle h \rangle$, because

$$h \quad = \quad j(k'm+1)h \quad = \quad jh' \quad \in \quad \langle h' \rangle\,.$$

Moreover we obtain $mod\ \mathbb{Z}$

$$\begin{aligned} \Theta(h') \quad &\equiv \quad (k'm+1)/m \quad \equiv \quad 1/m, \quad \text{and} \\ \tilde{\Theta}(h') \quad &\equiv \quad (k'm+1)/(2m) \quad \equiv \quad \frac{k'}{2} + \frac{1}{2m} \quad \equiv \quad \frac{1}{2m} + \frac{1}{2}. \end{aligned}$$

This proves (5.14.$a, b$).

Vice versa, we may define

$$h \quad := \quad jh'\,,$$

to turn a flip-flop doubling for $h'$ into a flop doubling for $h$. This completes the proof. □

Loosely speaking, lemma 5.12 tells us that flop doubling and flip-flop doubling are "stroboscopically equivalent": when observed with a suitable stroboscopic flashing, one bifurcation can be transformed into the other. So far we have favored the standard construction of Poincaré maps in our analysis of secondary bifurcations. In [Van4], Vanderbauwhede employs the operator setting (1.23), (1.24),

$$F(f,p,\lambda,\xi) \quad = \quad -\frac{1}{p}\dot{\xi} + f(\lambda,\xi) \quad = \quad 0\,, \tag{1.23}$$

instead. As in (1.26), symmetry comes in as the isotropy $H^{\Theta} \leq G \times S^1$ of periodic solutions $\xi \in \tilde{C}^1$ with (minimal) period 1. Reparametrizing $\xi$ by $\xi_k(\tau) := \xi(k\tau)$ one obtains an equivalent solution $\xi_k$ of (1.23) with related isotropy. This reparametrization lends itself to investigating e.g. period doubling bifurcations. Using an abstract Poincaré type map, all bifurcations of table 5.2 are then recovered (except of the freezing). In fact these bifurcations can all be viewed simply as $\mathbb{Z}_2$ symmetry breakings of $H^{\Theta}$. Moreover, the equivalent cases of flop doubling and flip-flop doubling are not distinguished by this approach.

However, it is not attempted in [Van4] to prove genericity in the sense of theorem 5.11. Rather, it is assumed a priori that the nontrivial eigenspace of the critical eigenvalue is one-dimensional. This gives rise to the $\mathbb{Z}_2$ symmetry breakings. We follow the more classically minded Poincaré map approach, mainly because we are lacking a perturbation theory to apply transversality theory directly to the operator setting (1.23). The obstacles might not be hard to overcome. But probably, measuring the total difficulty, the two approaches would provide equal variants.

We finish this section with an easy corollary, at last. Recall from §1.2, (1.28), that $\Theta$ and $\tilde{\Theta}$ can be represented by integers. We relate these integers for those cases of theorem 5.11 where $H = \tilde{H}$, using the same isomorphism $\iota$ from $H$ to the cyclic group of order $|H|$ in both cases.

**5.13 Corollary :**

*Under the assumptions of theorem 5.11, let $(\lambda_0, x_0)$ be a secondary bifurcation with $\tilde{H} = H$. Let $d \in D(|H|)$ denote the binary orbit of $\Theta$, cf. definition 2.4. Then*

$$\tilde{\Theta} \quad \in \quad -d,$$

*if $(\lambda_0, x_0)$ is a freezing. In all other cases*

$$\tilde{\Theta} \quad \in \quad d,$$

*i.e. $\Theta$ and $\tilde{\Theta}$ belong to the same binary orbit.*

**Proof :**

The cases $G = H = \mathbb{Z}_\infty$ can be read off trivially from table 5.2. The freezing is also covered there. Case c.1.1, the turn, is likewise trivial because $\tilde{\Theta} = \Theta$. In the remaining cases we note that

$$2\tilde{\Theta}(h) \quad \equiv \quad 1/m \quad = \quad \Theta(h) \qquad\qquad (mod\ \mathbb{Z}),$$

and therefore

$$2\tilde{\Theta} \quad \equiv \quad \Theta \qquad\qquad (mod\ |H|). \tag{5.15}$$

This proves the corollary. □

# 6 Generic global theory

## 6.1 Outline

With the local generic results (theorem 5.11, corollary 5.13) of the last section at hand, we can now aim at a proof of our main generic global results for $G = \mathbb{Z}_n$, $n \leq \infty$: a proof of theorems 2.6 and 2.7. To prove these theorems some information about generic global bifurcation diagrams is condensed into an equivariant orbit index $\Phi$. For $\Phi$ in the finite cyclic case $n < \infty$ see §6.2, definition 6.1. For $G = \mathbb{Z}_\infty$ see §6.6, definition 6.4. The orbit index $\Phi$ indicates which branch a global continuum should follow at secondary pitchfork and doubling bifurcations. The index $\Phi$ is a homotopy invariant, quite analogously to the case of no symmetry which was discussed in §3. We recall that the case of rotating waves, $G = \mathbb{Z}_\infty$, is somewhat special because we may continue rotating waves through a freezing, cf. table 5.2 and theorem 2.6. Therefore we discuss the case of finite cyclic symmetry $G = \mathbb{Z}_n$ first. In §6.3, proposition 6.2, we prove homotopy invariance of the orbit index $\Phi$, and in §6.4, proposition 6.3, we relate $\Phi$ to the center index $\phi^r$ at a generic center. These propositions then enable us in §6.5 to prove theorem 2.7, and a "grasshopping" reader may read this proof immediately, skipping the proofs of propositions 6.2 and 6.3 – at first. In §6.6, proposition 6.2 accounts for homotopy invariance of $\Phi$ through freezing and we conclude this chapter with a proof of theorem 2.6.

Below, our definitions of the orbit index $\Phi$ may appear to come out of the blue. Why should this particular index work to prove any global results? Couldn't we come up with a "better" index, tomorrow? Admittedly, our generic local results could yield other global results as well, e.g. somewhat more topologically minded ones. For a brief discussion see §9.3. Here we favor the following properties of the envisioned global bifurcating continuum $Z$:

$H = G = \mathbb{Z}_n$ remains fixed along $Z$; (6.1.a)

once $Z$ remains bounded, then the **minimal** period blows up – not just **any** period; (6.1.b)

$\aleph \neq 0$ leads to global bifurcation, not just odd $\aleph$; (6.1.c)

cf. theorems 2.6 and 2.7. Our definition of the orbit index $\Phi$ is the result of staring at the local bifurcation list (table 5.2) long enough to extract these global properties.

Throughout this section, $G = \mathbb{Z}_n$ is a (finite or infinite) cyclic group and $f$, satisfying (2.13), is generic $G$-equivariant so that theorem 5.11 holds.

## 6.2 The orbit index

To define the orbit index $\Phi$ we switch from the Poincaré map point of view, adopted in §5, to the operator setting (1.23) with (minimal) period as an explicit parameter. Recall that

$$F(f, \lambda, \xi(\cdot), p)(\tau) \quad = \quad -\frac{1}{p}\dot{\xi}(\tau) + f(\lambda, \xi(\tau)) \quad = \quad 0 \tag{6.2.a}$$

with $\xi$ of (minimal) period 1 iff

$$x(t) \quad := \quad \xi(t/p) \tag{6.2.b}$$

solves

$$\dot{x}(t) \quad = \quad f(\lambda, x(t)) \tag{1.1}$$

with (minimal) period $p$. Further, eigenvalues $\eta \in \mathbb{C}$ of the linearization of (6.2.$a$) are called Floquet exponents of $\xi(\cdot)$; their eigenfunctions $y(\tau)$ satisfy

$$(D_\xi F \cdot y)(\tau) \quad = \quad -\frac{1}{p}\dot{y}(\tau) + D_x f(\lambda, \xi(\tau))y(\tau) \quad = \quad \eta y(\tau)\,. \tag{6.3}$$

Note that $y$ is required to have period 1, here. Floquet multipliers as used in §3.3 relate to Floquet exponents by exponentiation: $\mu > 0$ is a Floquet multiplier of $x$ iff $\eta$ is a real Floquet exponent of $\xi$ and

$$\mu \quad = \quad e^{p\eta} \tag{6.4}$$

with the same algebraic multiplicities; for more details see e.g. [Ab&Mars, Hart]. So far the classical Floquet theory.

Period doubling bifurcations arise from real multipliers $\mu$ near $-1$, i.e. from **complex** exponents $\eta$. The following observation relates negative $\mu$ to **real** $\eta$. For $a \in \{+,-\}$ let

$$\mathcal{E}^a \quad := \quad \{y \in C^0(\mathbb{R}, X) \mid y(\tau+1) = ay(\tau)\}\,. \tag{6.5}$$

So $\mathcal{E}^+$ are the periodic functions with period 1, and $y \in \mathcal{E}^-$ are "anti-periodic", but periodic with period 2. Note that

$$D_\xi F: \quad \mathcal{E}^a \to \mathcal{E}^a \tag{6.6}$$

is an unbounded Fredholm operator with domain $\mathcal{E}^a \cap C^1(\mathbb{R}, X)$ and with compact resolvent. Moreover, $\mu < 0$ is a Floquet multiplier of $x$ iff

$$\mu \quad = \quad -e^{p\eta} \tag{6.7}$$

and $\eta$ is a real eigenvalue of $D_\xi F$ on $\mathcal{E}^-$. Again, the algebraic multiplicities of $\mu$ and $\eta$ coincide.

Remember our interest in periodic solutions $x(t)$ with symmetry $(H, K, \Theta)$, $H = G = \mathbb{Z}_n$. We restrict our attention to finite $n$, for a while. As in §5.4, (5.11.$a,b$) let $h \in H$ generate $H$ such that

$$\Theta(h) \quad = \quad \frac{1}{m}\,,$$

where $m = |H/K|$. In particular $h_0 := mh$ generates $K$. For $a, b \in \{+,-\}$ we define the following subspaces of $\mathcal{E}^a$. Given $H, \Theta$ let

$\mathcal{E}^a_b$: the set of $y \in C^0(\mathbb{R}, X)$ such that

$$y(\tau+1) \quad = \quad ay(\tau)\,, \quad \text{and} \tag{6.8.a}$$

$$\rho(h)y(\tau) \quad = \quad by(\tau + \Theta(h))\,. \tag{6.8.b}$$

A direct calculation shows that $\mathcal{E}^a_b$ does not depend on the particular choice of $h \in H$ which satisfies (5.11.$a,b$). We omit the details. We claim that by restriction

$$D_\xi F: \quad \mathcal{E}^a_b \to \mathcal{E}^a_b\,. \tag{6.9}$$

Indeed, $H^{\Theta}\xi = \xi$ and $y \in \mathcal{E}_b^a$ imply

$$\begin{aligned}
\rho(h)(D_\xi F \cdot y)(\tau) &= -\frac{1}{p}\rho(h)\dot{y}(\tau) + \rho(h)D_x f(\lambda, \xi(\tau))y(\tau) = \\
&= -b\frac{1}{p}\dot{y}(\tau + \Theta(h)) + D_x f(\lambda, \rho(h)\xi(\tau))\rho(h)y(\tau) = \\
&= b\Big(-\frac{1}{p}\dot{y}(\tau + \Theta(h)) + D_x f(\lambda, \xi(\tau + \Theta(h)))\; y(\tau + \Theta(h))\Big) = \\
&= b(D_\xi F \cdot y)(\tau + \Theta(h)),
\end{aligned}$$

hence $D_\xi F \cdot y \in \mathcal{E}_b^a$. This allows us to define

$$\sigma_b^a : \quad \text{the number of positive real eigenvalues } \eta \text{ of } D_\xi F \text{ on } \mathcal{E}_b^a, \text{ counting algebraic multiplicities;} \tag{6.10}$$

cf.(6.2), (6.3), (6.8), (6.9). Note that $\sigma_b^a$ is finite because positive real eigenvalues of $D_\xi F$ on $\mathcal{E}^a$ relate to real Floquet multipliers $\mu$ with $a\mu > 1$, by (6.4) and (6.7).

With these preparations we can now define the orbit index $\Phi$ for finite cyclic groups.

**6.1 Definition :**

*Consider $G = \mathbb{Z}_n$, $n < \infty$, and let $(\lambda, x(t))$ be a periodic solution with symmetry $(H, K, \Theta)$, rescaled to $\xi$ with minimal period 1. If $H = G$, then we define the* **orbit index** *$\Phi$ of $x$ to be*

$$\Phi(\lambda, x) := \begin{cases} (-1)^{\sigma_+^+} & \text{if } \sigma_+^- \text{ and } \sigma_-^- \text{ are both even,} \\ -(-1)^{\sigma_+^+} & \text{if } \sigma_+^- \text{ and } \sigma_-^- \text{ are both odd,} \\ 0 & \text{otherwise,} \end{cases} \tag{6.11}$$

*where the $\sigma_b^a$ are defined in (6.10) above. In other words,*

$$\Phi(\lambda, x) = \frac{1}{2}\left((-1)^{\sigma_+^+ + \sigma_+^-} + (-1)^{\sigma_+^+ + \sigma_-^-}\right). \tag{6.11$'$}$$

*If $H < G$ we put $\Phi(\lambda, x) := 0$.*

For the case of no symmetry, i.e. $G = \{0\}$ as discussed in §3, the orbit index $\Phi$ reduces to the orbit index of Mallet-Paret & Yorke [M-P&Y2] introduced in §3.4, (3.11). Indeed, $\mathcal{E}_+^a = \mathcal{E}^a$ and $\mathcal{E}_-^a = \{0\}$, by definition (6.8.$a$, $b$), because $h = 0$, $\Theta(h) = 0$ in that case. Relating eigenvalues $\eta$ of $D_\xi F$ back to Floquet multipliers, by (6.4) and (6.7), this implies that $\sigma_+^+ = \sigma^+$ and $\sigma_+^- = \sigma^-$ whereas $\sigma_-^- = 0$. Thus (6.11)$'$ reduces to definition (3.11)$'$ of the orbit index $\Phi$ with no symmetry present.

## 6.3 Homotopy invariance

The next proposition formulates homotopy invariance of $\Phi$ for both finite and infinite cyclic groups $G$. Never mind that $\Phi$ is not yet defined for infinite $G$. With definition 6.4 of $\Phi$ for $G = \mathbb{Z}_\infty$ below, the proposition will remain valid and we just postpone the proof for that case.

**6.2 Proposition :**

*Let $G = \mathbb{Z}_n, n \leq \infty, f$ generic with a periodic solution $(\lambda_0, x_0(\cdot))$, and fix $\varepsilon > 0$ small enough.*

*Then for any $0 < \delta \leq \delta_0(\varepsilon)$*

$$\sum_{\lambda=\lambda_0-\delta} \Phi \quad = \quad \sum_{\lambda=\lambda_0+\delta} \Phi, \tag{6.12}$$

*where the sums range over all orbit indices $\Phi$ of periodic solutions $(\lambda, x(\cdot))$, with the indicated values of $\lambda$, which remain in an $\varepsilon$-neighborhood of $(\lambda_0, x_0(\cdot))$. Geometrically, these solutions lie on the branches emanating from $(\lambda_0, x_0(\cdot))$ to the left respectively right.*

**Proof ($n < \infty$) :**

We study all possible generic secondary bifurcations from $(\lambda_0, x_0(\cdot))$, case by case. Type 0, turn, and the two pitchforks are easy cases which we handle first. The doubling cases require some notational preparation and are tackled afterwards. We treat only flop doubling and flip doubling, since the third case of a flip-flop doubling is equivalent to a flop doubling by lemma 5.12. Indeed, interchanging $h$ and $h'$, there, does not affect the orbit index $\Phi$ because the spaces $\mathcal{E}_b^a$ remain unchanged. Throughout we may assume $H = G$ for the symmetry $(H, K, \Theta)$ of $(\lambda_0, x_0(\cdot))$. Otherwise $\Phi = 0$, locally, and the proposition holds trivially.

If $(\lambda_0, x_0(\cdot))$ is **type 0** then the $\sigma_b^a$ remain invariant *mod* 2 as $\lambda$ increases through $\lambda_0$. Indeed, $\sigma_b^a$ cannot change by eigenvalues of $D_\xi F$ on $\mathcal{E}_b^a$ passing through zero, because $(\lambda_0, x_0)$ is type 0. Thus $\sigma_b^a$ can change only by conjugate complex pairs of eigenvalues of $D_\xi F$ on $\mathcal{E}_b^a$ becoming real, or vice versa. This does not affect $\sigma_b^a$ *mod* 2. Hence $\Phi$ is the same to the left and to the right of the bifurcation orbit and (6.12) holds.

If $(\lambda_0, x_0(\cdot))$ is a **turn**, then we find a corresponding branch $(\lambda(s), \xi(s, \cdot), p(s))$ of solutions of

$$F(f, \lambda(s), \xi(s, \cdot), p(s)) \quad = \quad 0$$

with $\lambda(0) = \lambda_0, \xi(0, t/p(0)) = x_0(t)$, such that

$$D_s\lambda(0) = 0 \qquad \text{and} \qquad y_0 := D_s\xi(0, \cdot) \;\notin\; span\; \dot{\xi}(0, \cdot)\,,$$

where $\dot{\xi}$ denotes the derivative with respect to time $\tau$. In particular, differentiation with respect to $s$ yields

$$D_\xi F(f, \lambda(0), \xi(0, \cdot), p(0)) y_0 \quad = \quad c\dot{\xi}(0, \cdot)$$

for some real constant $c$, since $D_p F$ is a multiple of $\dot{\xi}$. Moreover, $H^\Theta \xi(s, \cdot) = \xi(s, \cdot)$ implies

$$y_0 \quad \in \quad \mathcal{E}_+^+\,.$$

Thus $y_0 \in \mathcal{E}_+^+$ is a generalized eigenfunction of $D_\xi F$ for the Floquet exponent 0, i.e. for the (nontrivial) critical Floquet multiplier $+1$ at $(\lambda_0, x_0)$. Following the local branch through $(\lambda_0, x_0)$, the transverse crossing condition (3.7.*b*) for the critical Floquet multiplier implies that $\sigma_+^+$ changes by 1 at $(\lambda_0, x_0)$. By condition (3.7.*a*)$'$, all other $\sigma_b^a$ remain unchanged. Thus the indices $\Phi$ have opposite sign at the two periodic solutions for $\lambda$ on the one side of $\lambda_0$. The sum of $\Phi$ for $\lambda$ on the other side of $\lambda_0$ is empty (zero). Again (6.12) holds.

If $(\lambda_0, x_0(\cdot))$ is any of the **pitchfork** cases (flip or flip-flop) then $\tilde{H} < H = G$ on the secondary branches, by table 5.2. This implies $\Phi = 0$ on the secondary branches. It is therefore sufficient to show that $\Phi$ remains unchanged along the primary branch. Indeed, one eigenfunction $y_0(\tau)$ for the critical Floquet multiplier +1 lies in $\mathcal{E}_-^+$: by virtual symmetry, proposition 4.3 and corollary 4.6, there exists $y_0 \in \mathcal{E}^+ \cap \textit{ker } D_\xi F$ such that the pair $(\xi_0, y_0)$ has symmetry $(\tilde{H}, \tilde{K}, \tilde{\Theta})$, cf. table 5.2. In particular, $y_0 \notin \textit{span } \dot{\xi}_0$ and

$$\rho(h)y_0 \quad \neq \quad y_0(\Theta(h)).$$

Because the critical nontrivial Floquet multiplier +1 is simple at the pitchforks, we can conclude

$$\rho(h)y_0 \quad = \quad -y_0(\Theta(h))$$

from equivariance of $D_\xi F$ with respect to $H^\Theta$. Thus $y_0 \in \mathcal{E}_-^+$ as claimed. Therefore $\Phi$ remains unchanged along the primary branch because, *mod* 2, only $\sigma_-^+$ changes and $\sigma_-^+$ is irrelevant for $\Phi$.

To complete the proof we have to consider the doubling cases (flop and flip). Their treatment requires some more preparation. For definiteness, we consider the geometric situation of figure 5.7, first, assuming that the secondary branch $(\lambda(s), x(s))$ bifurcates to the right of $\lambda_0$, i.e. $\lambda(s) > \lambda_0$. Note that the critical multiplier $\mu$ near $-1$ on the primary branch $x_\lambda$ yields an eigenfunction of $D_\xi F(\lambda, \xi_\lambda)$ in $\mathcal{E}_+^-$ if $(\lambda_0, x_0)$ is a flop doubling. Again, we use virtual symmetry and table 5.2 here. Then only $\sigma_+^-$ changes by 1, while all other $\sigma_b^a$ remain unchanged, as $\lambda$ increases through $\lambda_0$. In contrast, at a flip doubling $\sigma_-^-$ changes by 1 and all other $\sigma_b^a$ remain unchanged. In any case, $\Phi(x_\lambda) = 0$ for $\lambda$ on one side of $\lambda_0$, while $\Phi(x_\lambda) \neq 0$ on the other side. Let us assume for the moment that $\Phi(x_\lambda) \neq 0$ for $\lambda < \lambda_0$ near $\lambda_0$, again for definiteness. We postpone comments on the remaining cases to the end of the proof.

Let $\xi_\lambda$ resp. $\tilde{\xi}_\lambda$ denote the rescaled periodic solutions $x_\lambda$ resp. $x(s), \lambda = \lambda(s)$ on the primary resp. secondary branch with minimal period normalized to one. See (6.2.$b$) and fig. 6.1.

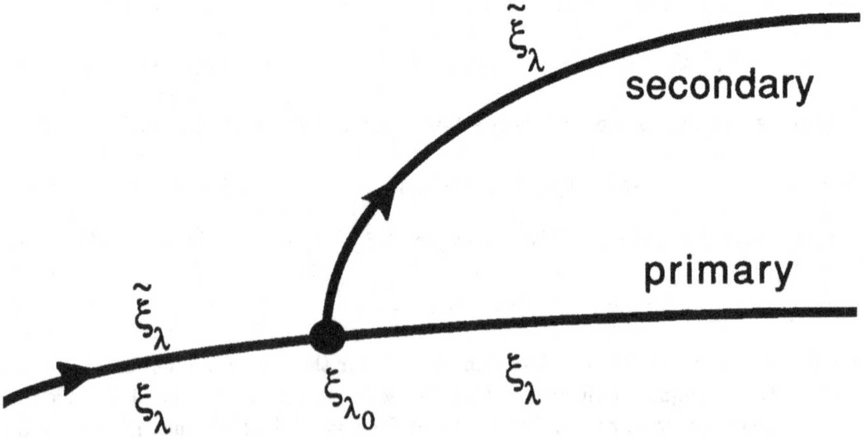


**Fig. 6.1 Period doubling bifurcations**

We would like to discuss eigenvalues of

$$D_\xi F(\lambda,\xi) \quad = \quad -\frac{1}{p}\frac{d}{d\tau} + D_x f(\lambda,\xi_\lambda)$$

following $\xi = \xi_\lambda$ for $\lambda \nearrow \lambda_0$ on the primary branch and then $\xi = \tilde{\xi}_\lambda$ along the secondary branch. Unfortunately $p$ jumps to $\tilde{p} = 2p$ at $(\lambda_0, x_{\lambda_0})$.

As a remedy we lift $\xi_\lambda$ , $\lambda \leq \lambda_0$, to an orbit $\tilde{\xi}_\lambda$ with non-minimal period 1 and minimal period 1/2, defining

$$\tilde{\xi}_\lambda(\tau) \quad := \quad \xi_\lambda(2\tau) \qquad\qquad \text{for } \lambda \leq \lambda_0 . \tag{6.13}$$

By this rescaling, $p$ gets replaced by $\tilde{p} := 2p$ along the primary branch and

$$\lambda \quad \mapsto \quad D_\xi \tilde{F}(\lambda,\tilde{\xi}_\lambda) := -\frac{1}{\tilde{p}}\frac{d}{d\tau} + D_x f(\lambda,\tilde{\xi}_\lambda) \tag{6.14}$$

now depends continuously (but not differentiably) on $\lambda$ near $\lambda_0$ in the uniform operator topology of $\mathcal{L}(\tilde{C}^1,\tilde{C}^0)$.

To analyse eigenvalues of $D_\xi F$ (leading to $\Phi(x_{\lambda_0-\delta})$) as well as of $D_\xi \tilde{F}$ (leading to $\Phi(x(s)), \lambda(s) = \lambda_0 + \delta$) we introduce two more spaces which we call ${}^2\mathcal{E}^a_b$ and $\tilde{\mathcal{E}}^a_b$. Let

$$\begin{array}{ll} {}^2\mathcal{E}^a_b : & \text{the set of all } \tilde{y} \in C^0(\boldsymbol{R},X) \text{ for which there exists} \\ & \text{some } y \in \mathcal{E}^a_b \text{ such that } \tilde{y}(\tau) = y(2\tau)\text{, for all } \tau \in \boldsymbol{R}, \\ \tilde{\mathcal{E}}^a_b : & \text{the space defined analogously to } \mathcal{E}^a_b \text{ but using the} \\ & \text{symmetry } (\tilde{H},\tilde{K},\tilde{\Theta}) \text{ of the secondary branch instead} \\ & \text{of } (H,K,\Theta). \end{array} \tag{6.15}$$

Thus ${}^2\mathcal{E}^a_b$ is the lifted space $\mathcal{E}^a_b$ whereas $\tilde{\mathcal{E}}^a_b$ is the space "$\mathcal{E}^a_b$" associated to the secondary symmetry. Likewise let $\mathcal{E}^a_b(\eta)$ denote the lifted generalized eigenspace $\mathcal{E}^a_b(\eta)$ of a Floquet exponent $\eta$, and $\tilde{\mathcal{E}}^a_b(\eta)$ the generalized eigenspace of $\eta$ as an eigenvalue of $D_\xi\tilde{F}(\lambda,\tilde{\xi}_\lambda)$ on $\tilde{\mathcal{E}}^a_b$. Obviously

$$dim\ {}^2\mathcal{E}^a_b(\eta) \quad = \quad dim\ \mathcal{E}^a_b(\eta)\,,$$

for any $\eta$. Note that the spaces $\mathcal{E}^a_b(\eta), {}^2\mathcal{E}^a_b(\eta), \tilde{\mathcal{E}}^a_b(\eta)$ depend on $\lambda$ because they refer to $\xi_\lambda, \xi_\lambda, \tilde{\xi}_\lambda$, respectively.

It is a first benefit of these spaces that an exchange-of-stability formula holds if we follow the lifted branch $\tilde{\xi}_\lambda$ through the bifurcation point $\tilde{\xi}_{\lambda_0}$. More precisely

$$\tilde{\sigma}^a_b \quad := \quad dim \bigoplus_{\eta>0} \tilde{\mathcal{E}}^a_b(\eta) \quad = \quad \sum_{\eta>0} dim\ \tilde{\mathcal{E}}^a_b(\eta) \tag{6.16}$$

remains invariant as $\lambda$ decreases through $\lambda_0$. Because we have assumed that $\Phi(x_\lambda) = 0$ for $\lambda > \lambda_0$ we have therefore proved homotopy invariance of $\Phi$ if we can only show

$$\begin{aligned} \Phi(x_{\lambda_0-\delta}) \quad :&= \quad \frac{1}{2}\left((-1)^{\sigma^+_+ + \sigma^-_+} + (-1)^{\sigma^+_+ + \sigma^-_-}\right) \quad = \\ &= \quad \frac{1}{2}\left((-1)^{\tilde{\sigma}^+_+ + \tilde{\sigma}^-_+} + (-1)^{\tilde{\sigma}^+_+ + \tilde{\sigma}^-_-}\right) \end{aligned} \tag{6.17}$$

at $\lambda = \lambda_0 - \delta$. To achieve this we claim that for any choice of $a, b \in \{+,-\}$ and $\eta \in \mathbb{R}$

$$dim\ \tilde{\mathcal{E}}_b^a(\eta) \equiv dim \left( \tilde{\mathcal{E}}_b^a(\eta) \cap \Big( \bigoplus_{a',b' \in \{+,-\}} {}^2\mathcal{E}_{b'}^{a'} \Big) \right) \tag{6.18.a}$$

$$\tilde{\mathcal{E}}_b^a(\eta) \cap \left( \bigoplus_{a',b' \in \{+,-\}} {}^2\mathcal{E}_{b'}^{a'} \right) = \begin{cases} {}^2\mathcal{E}_b^+(\eta) \oplus {}^2\mathcal{E}_{\pm b}^-(\eta) & \text{if } a = +\,, \\ 0 & \text{if } a = -\,, \end{cases} \tag{6.18.b}$$

where the sign $\pm$ in ${}^2\mathcal{E}_{\pm b}^-(\eta)$ is + for a flop and – for a flip. The congruence $\equiv$ here and below is understood *mod* 2. Accepting both claims for the moment, we can prove (6.17) as follows. Summing over $\eta > 0$ in (6.18.$a, b$) we obtain from the above

$$\begin{aligned} \tilde{\sigma}_+^+ &= \sum_{\eta>0} dim\ \tilde{\mathcal{E}}_+^+(\eta) \equiv \sum_{\eta>0} dim\ {}^2\mathcal{E}_+^+(\eta) + dim\ {}^2\mathcal{E}_\pm^-(\eta) \\ &= \sum_{\eta>0} dim\ \mathcal{E}_+^+(\eta) + dim\ \mathcal{E}_\pm^-(\eta) = \sigma_+^+ + \sigma_\pm^-, \quad \text{and} \\ \tilde{\sigma}_+^- &\equiv \tilde{\sigma}_-^- \equiv 0\,. \end{aligned}$$

Because $\Phi(x_{\lambda_0-\delta}) \neq 0$ we also know

$$\sigma_+^- \equiv \sigma_-^-\,.$$

Together, this implies (6.17) because

$$\tilde{\sigma}_+^+ + \tilde{\sigma}_b^- \equiv \tilde{\sigma}_+^+ \equiv \sigma_+^+ + \sigma_+^- \equiv \sigma_+^+ + \sigma_-^-$$

for any $b \in \{+,-\}$. It only remains to prove (6.18.$a, b$).

To prove (6.18.$a$) we lift the cylic group $H^\Theta = \{(h, \Theta(h)) | h \in H\} \leq G \times S^1$ to a covering group

$$\tilde{H}^\Theta := \{(h, \vartheta) \mid 2\vartheta \equiv \Theta(h) \pmod{\mathbb{Z}}\} \leq G \times S^1\,.$$

Because we consider $\xi_\lambda, \tilde{\xi}_\lambda$ with $\lambda < \lambda_0$, the linearization $D_\xi \tilde{F}(\lambda, \tilde{\xi}_\lambda)$ commutes with the action of $\tilde{H}^\Theta$. Indeed

$$\begin{aligned} \tilde{\rho}(h,\vartheta)\,(D_\xi \tilde{F}(\lambda, \tilde{\xi}_\lambda) \cdot \tilde{y})\,(\tau) &= -\frac{1}{\tilde{p}}\frac{d}{d\tau}\rho(h)\tilde{y}(\tau - \vartheta) + D_x f(\lambda, \rho(h)\tilde{\xi}_\lambda(\tau - \vartheta))\,\rho(h)\tilde{y}(\tau - \vartheta) \\ &= D_\xi \tilde{F}(\lambda, \tilde{\xi}_\lambda) \cdot (\tilde{\rho}(h,\vartheta)\tilde{y})\,(\tau) \end{aligned}$$

because $\tilde{\xi}_\lambda(\tau) = \xi_\lambda(2\tau)$ implies

$$\rho(h)\tilde{\xi}_\lambda(\tau - \vartheta) = \rho(h)\xi_\lambda(2\tau - \Theta(h)) = \xi_\lambda(2\tau) = \tilde{\xi}_\lambda(\tau)\,.$$

Because $\tilde{H}^\Theta$ thus commutes with $D_\xi \tilde{F}(\lambda, \tilde{\xi}_\lambda)$, the eigenspaces $\tilde{\mathcal{E}}_b^a(\eta)$ are all invariant under the action of $\tilde{H}^\Theta$. The irreducible representations of $\tilde{H}^\Theta$ all have real dimension either one or two because $\tilde{H}^\Theta \leq \mathbb{Z}_n \times S^1$ is compact abelian, cf. [Brö&tD, §II.8]. The representation spaces corresponding to real one-dimensional representations of $\tilde{H}^{\Theta}$ are exactly the spaces ${}^2\mathcal{E}_{b'}^{a'}$ defined in (6.15) above. Indeed, $a'$ accounts for the action of $(0, \frac{1}{2}) \in \tilde{H}^{\Theta}$

$$\tilde{y}(\tau + \frac{1}{2}) \quad = \quad a'\tilde{y}(\tau)\,;$$

whereas $b'$ accounts for the action of $(h, \frac{1}{2m})$

$$\tilde{\rho}(h, \frac{1}{2m})\tilde{y}(\tau) \quad = \quad \rho(h)y(2\tau - \frac{1}{m}) \quad = \quad b'y(2\tau) \quad = \quad b'\tilde{y}(\tau).$$

Because all other irreducible representations are two-dimensional, decomposition into irreducible representations implies (6.18.$a$).

We prove (6.18.$b$) next, separately for a flop and a flip doubling at $(\lambda_0, x_0)$.

**Case 1, flop doubling :**

Let $\tilde{y} \in \tilde{\mathcal{E}}_b^a(\eta) \cap {}^2\mathcal{E}_{b'}^{a'}$. By table 5.2 we have $\tilde{h} = h, \tilde{\Theta}(\tilde{h}) = 1/\tilde{m} = 1/2m$. Thus

$$\begin{aligned} \tilde{y}(\tau + 1) &= a\tilde{y}(\tau)\,, \\ \rho(\tilde{h})\tilde{y}(\tau) &= b\tilde{y}(\tau + \tilde{\Theta}(\tilde{h})) \quad = \quad b\tilde{y}(\tau + \frac{1}{2m})\,, \end{aligned}$$

by $\tilde{y} \in \tilde{\mathcal{E}}_b^a(\eta)$. On the other hand $\tilde{y} \in {}^2\mathcal{E}_{b'}^{a'}$, i.e. $\tilde{y}(\tau) = y(2\tau)$, $y \in \mathcal{E}_{b'}^{a'}$, implies

$$\begin{aligned} \tilde{y}(\tau + \frac{1}{2}) &= a'\tilde{y}(\tau)\,, \\ \rho(\tilde{h})\tilde{y}(\tau) &= \rho(h)\tilde{y}(\tau) \quad = \quad \rho(h)y(2\tau) \quad = \\ &= b'y(2\tau + \Theta(h)) \quad = \quad b'y(2\tau + \frac{1}{m}) \quad = \quad b'\tilde{y}(\tau + \frac{1}{2m})\,. \end{aligned}$$

Comparing the two calculations we immediately observe that $\tilde{\mathcal{E}}_b^a(\eta) \cap {}^2\mathcal{E}_{b'}^{a'} = \{0\}$ unless $a = (a')^2 = +$ and $b = b'$. If $a = +$, $b = b'$, then $y$ is a generalized eigenfunction of $D_\xi F(\lambda, \xi_\lambda)$ in $\mathcal{E}_{b'}^{a'}$ with eigenvalue $\eta$, by the usual rescaling argument. Vice versa, any such eigenfunction $y$ rescales to an element $\tilde{y} \in \tilde{\mathcal{E}}_b^a(\eta) \cap {}^2\mathcal{E}_{b'}^{a'}$ in that case. Summarizing, we obtain

$$\tilde{\mathcal{E}}_b^a(\eta) \cap {}^2\mathcal{E}_{b'}^{a'} \quad = \quad \begin{cases} {}^2\mathcal{E}_b^{a'}(\eta) & \text{if } a = +\,,\ b' = b \\ \{0\} & \text{otherwise}\,. \end{cases}$$

This proves (6.18.$b$).

**Case 2, flip doubling :**

This time $\tilde{h} = (k'm + 2)h$, where $k'm$ is odd, and $\tilde{\Theta}(\tilde{h}) = 1/\tilde{m} = 1/m$. As in the previous case, $\tilde{y} \in \tilde{\mathcal{E}}_b^a(\eta)$ implies

$$\begin{aligned} \tilde{y}(\tau + 1) &= a\tilde{y}(\tau) \\ \rho(\tilde{h})\tilde{y}(\tau) &= b\tilde{y}(\tau + \tilde{\Theta}(\tilde{h})) \quad = \quad b\tilde{y}(\tau + \frac{1}{m})\,. \end{aligned}$$

On the other hand, $\tilde{y} \in {}^2\mathcal{E}_{b'}^{a'}$ implies

$$\tilde{y}(\tau + \frac{1}{2}) \quad = \quad a'\tilde{y}(\tau)\,, \qquad \text{and}$$

$$\begin{aligned}\rho(\tilde h)\tilde y(\tau) &= \rho(h)^{k'm+2}\,y(2\tau) = \\ &= (b')^{k'm}y(2\tau+(k'm+2)\Theta(h)) = \\ &= b'\tilde y(\tau+\frac{k'm+2}{2m}) = b'\tilde y(\tau+\frac{1}{m}+\frac{k'}{2}) = \\ &= a'b'\tilde y(\tau+\frac{1}{m})\,.\end{aligned}$$

Thus we conclude this time that

$$\tilde{\mathcal{E}}^a_b(\eta)\cap{}^2\mathcal{E}^{a'}_{b'} = \begin{cases} {}^2\mathcal{E}^{a'}_b(\eta) & \text{if } a=+\,,\ b=a'b' \\ \{0\} & \text{otherwise}\,. \end{cases}$$

Again, this proves (6.18.$b$).

The above considerations prove homotopy invariance of the index $\Phi$ assuming bifurcation to the right of $\lambda_0$ and, in the doubling cases, assuming also $\Phi(\lambda,x_\lambda)\neq 0$ for $\lambda<\lambda_0$. Replacing $\lambda$ by $-\lambda$, if necessary, it is actually sufficient to prove homotopy invariance only for those remaining doubling bifurcations where $\Phi(\lambda,x_\lambda)=0$ for $\lambda<\lambda_0$ near $\lambda_0$ and bifurcation still occurs to the right, as before. As we have noted before, this implies $\Phi(\lambda,x_\lambda)\neq 0$ for $\lambda>\lambda_0$, because only $\sigma^-_+$ resp. $\sigma^-_-$ changes at a flop resp. flip doubling while all other $\sigma^a_b$ remain unchanged. In short hand: $\sigma^-_\pm$ changes by 1 while e.g. $\sigma^+_+,\sigma^-_\mp$ remain unchanged. Again, the sign $\pm$ is $+$ for a flop and $-$ for a flip. Evaluating all $\sigma^a_b$ at $(\lambda_0-\delta,\xi_{\lambda_0-\delta})$ we thus conclude from $\Phi(\lambda_0+\delta,x_{\lambda_0+\delta})\neq 0$ that

$$\Phi(\lambda_0+\delta,x_{\lambda_0+\delta}) = (-1)^{\sigma^+_+ +\sigma^-_\mp}\,. \tag{6.19.a}$$

Evaluating all $\tilde\sigma^a_b$ at $(\lambda_0-\delta,\tilde\xi_{\lambda_0-\delta})$ we obtain from exchange of stability that

$$\Phi\,(\lambda_0+\delta,x(s)) = \frac{1}{2}\left((-1)^{\tilde\sigma^+_+ +\tilde\sigma^-_+} + (-1)^{\tilde\sigma^+_+ +\tilde\sigma^-_-}\right), \tag{6.19.b}$$

as before, with $\lambda_0+\delta=\lambda(s)$. Because (6.18.$a,b$) remain valid

$$\begin{aligned}\tilde\sigma^+_+ &\equiv \sigma^+_+ + \sigma^-_\pm\,, \quad \text{and} \\ \tilde\sigma^-_+ &\equiv \tilde\sigma^-_- \equiv 0\,.\end{aligned} \tag{6.19.c}$$

still hold. Also, $\Phi(\lambda_0-\delta,x_{\lambda_0-\delta})=0$ implies

$$\sigma^-_\pm \equiv \sigma^-_\mp + 1\,. \tag{6.19.d}$$

From (6.19.$a$-$d$) we finally conclude

$$\begin{aligned}\Phi(\lambda_0+\delta,x(s)) &= (-1)^{\tilde\sigma^+_+} = (-1)^{\sigma^+_+ +\sigma^-_\pm} = \\ &= -(-1)^{\sigma^+_+ +\sigma^-_\mp} = -\Phi(\lambda_0+\delta,x_{\lambda_0+\delta})\,.\end{aligned}$$

This completes the proof of homotopy invariance of $\Phi$ in all generic cases with finite cyclic symmetry. □

## 6.4 Orbit index and center index

The next proposition relates the orbit index $\Phi$, defined in (6.11), to the center index $\text{ф}^r$ at a generic center. Again the formulation is valid for both finite and infinite cyclic groups $G$ with $\Phi$ from definitions 6.1 and 6.4. But we postpone definition and proof of the case $G = \mathbb{Z}_\infty$ to §6.6.

**Proposition 6.3 :**
*Let $G = \mathbb{Z}_n$, $n \leq \infty$, $f$ generic with a generic $G$-center $(\lambda_0, x_0)$ and bifurcating local branch $(\lambda(s), x(s,\cdot))$ of periodic solutions with symmetry $(H, K, \Theta)$ and with representation $r = \Theta$ at the center $(\lambda_0, x_0)$ (cf. lemma 5.1).*
*Then the orbit index $\Phi$ satisfies*

$$\Phi(\lambda(s), x(s,\cdot)) \quad = \quad \text{ф}^r(\lambda_0, x_0) \cdot sign\,(\lambda(s) - \lambda_0) \quad \neq \quad 0, \tag{6.20}$$

*for any small enough positive $s$. Here $\text{ф}^r(\lambda_0, x_0)$ denotes the center index of $(\lambda_0, x_0)$ (see definition 2.3).*

**Proof ($n < \infty$) :**
We model the proof in technical analogy to the proof of homotopy invariance proposition 6.2. We consider the case $\lambda(s) > \lambda_0$ of bifurcation to the right, first, postponing $\lambda(s) < \lambda_0$ to the end of the proof. Rescaling the minimal period $p(s)$ of $(\lambda(s), x(s,\cdot))$ to 1 we obtain the branch $\tilde{\xi}_\lambda$ of bifurcating periodic solutions

$$\tilde{\xi}_\lambda(\tau) \quad = \quad x(s, \tau p(s)) \qquad \text{with} \quad \lambda = \lambda(s) > \lambda_0\,.$$

Again, we can extend this branch $\tilde{\xi}_\lambda \in C^0(\mathbb{R}, X)$ continuously to $\lambda \leq \lambda_0$ defining

$$\tilde{\xi}_\lambda(\tau) \quad := \quad x_\lambda \qquad\qquad \text{for } \lambda \leq \lambda_0 \text{ near } \lambda_0, \quad \tau \in \mathbb{R}\,,$$

where $(\lambda, x_\lambda)$ denotes the stationary branch through the generic center $(\lambda_0, x_0)$. The proof of proposition 6.3 will consist of a careful study of the eigenvalues $\eta$ of the linearization

$$D_\xi \tilde{F}(\lambda, \tilde{\xi}_\lambda) \quad := \quad -\frac{1}{\tilde{p}_\lambda}\frac{d}{d\tau} + D_x f(\lambda, \tilde{\xi}_\lambda) \tag{6.14}$$

as $\lambda$ increases through zero. Here we put

$$\tilde{p}_\lambda := \begin{cases} p(s) & \text{for } \lambda = \lambda(s) > \lambda_0\,, \\ 2\pi/\beta(\lambda) & \text{for } \lambda \leq \lambda_0\,, \end{cases}$$

where $\alpha(\lambda) \pm i\beta(\lambda)$ denote the eigenvalues of $D_x f(\lambda, x_\lambda)$ near $\pm i\beta(\lambda_0)$, as in (2.3), with eigenvector $y_\lambda$. Note that $D_\xi \tilde{F}(\lambda, \tilde{\xi}_\lambda)$ depends continuously on $\lambda$ near $\lambda_0$ in the uniform operator topology $\mathcal{L}(\tilde{C}^1, \tilde{C}^0)$.

To relate $\text{ф}$ and $\Phi$ we define the spaces $\mathcal{E}^a_b$ as in (6.8.$a, b$) with the symmetry $(H, K, \Theta) = (G, K, r)$ of the bifurcating branch; for $\Theta = r$ cf. lemma 5.1. We observe that the linearization (6.14) maps each space $\mathcal{E}^a_b$ into itself. Denoting the sum of multiplicities of eigenvalues $\eta > 0$

of $D_\xi\tilde{F}(\lambda,\tilde{\xi}_\lambda)$ in $\mathcal{E}_b^a$ by $\sigma_b^a=\sigma_b^a(\lambda)$, to emphasize their dependence on $(\lambda,\tilde{\xi}_\lambda)$, we claim that for all small enough positive $\delta$

$$\sigma_+^+(\lambda_0+\delta) \quad \equiv \quad \begin{cases} \sigma_+^+(\lambda_0-\delta) & \text{if } \chi^r=+1, \\ \sigma_+^+(\lambda_0-\delta)+1 & \text{if } \chi^r=-1, \end{cases} \tag{6.21.a}$$

$$\sigma_b^a(\lambda_0+\delta) \quad \equiv \quad \sigma_b^a(\lambda_0-\delta)\,, \quad \text{for all } (a,b)\neq(+,+) \tag{6.21.b}$$

Here $\chi^r$ is the crossing number and $\equiv$ indicates congruence *mod* 2. Accepting (6.21.*a*, *b*), for the moment, the relation (6.20) between $\mathit{\Phi}$ and $\Phi$ follows, by definitions 2.3 and 6.1 of $\Phi$ and $\mathit{\Phi}$, once we show that

$$\sigma_b^-(\lambda_0-\delta) \quad \equiv \quad 0 \qquad \text{for each } b\in\{+,-\}\,, \tag{6.22.a}$$

$$\sigma_+^+(\lambda_0-\delta) \quad \equiv \quad E^0(\lambda_0)\,. \tag{6.22.b}$$

We still remember, of course, that we consider the case $\lambda(s)-\lambda_0>0$. It thus remains to prove (6.21), (6.22), in that case.

We prove (6.22) first. Because $\tilde{\xi}_\lambda(\tau)$ is $G$-invariant and independent of $\tau$ for $\lambda=\lambda_0-\delta<\lambda_0$, the group $G\times S^1$ acts on each generalized eigenspace $\mathcal{E}_b^a(\eta)$ of $D_\xi\tilde{F}(\lambda,\tilde{\xi}_\lambda)$. In particular, $S^1$ acts by time shift. But the only irreducible representation of $S^1$ with real dimension 1 is the trivial representation, and all other irreducible representations have real dimension two. Thus the $\sigma_b^a$ can be obtained *mod* 2 from the corresponding multiplicities of the restriction of $D_\xi\tilde{F}(\lambda,\tilde{\xi}_\lambda)$ to the spaces

$$\mathcal{E}_b^a\cap X\,,$$

where $X\subseteq\tilde{C}^0(\mathbb{R},X)$ stands for the constant functions in $C^0(\mathbb{R},X)$. But obviously

$$\mathcal{E}_b^-\cap X \quad = \quad \{0\}\,,$$

by definition (6.8) of $\mathcal{E}_b^-$. This proves (6.22.*a*). It is equally obvious that

$$D_\xi\tilde{F}(\lambda,\tilde{\xi}_\lambda) \quad = \quad D_x f(\lambda,x_\lambda) \quad \text{on} \quad \mathcal{E}_+^+\cap X \;=\; X^G\,,$$

proving (6.22.*b*).

We prove (6.21) next. Real pairs of eigenvalues becoming complex conjugate, and vice versa, does not change any $\sigma_b^a$ (*mod* 2). To prove (6.21) we therefore need to discuss only the behavior of the critical (algebraically and geometrically) double Floquet exponent $\eta=0$ as $\lambda$ increases through $\lambda_0$. This Floquet exponent is related to the eigenfunctions associated to the purely imaginary eigenvalues $\pm i\beta(\lambda_0)$ at the generic center $(\lambda_0,x_0)$, cf. (2.4). By the proof of lemma 5.1, these eigenfunctions of $\eta=0$ belong to the space $\mathcal{E}_+^+$. This proves (6.21.*b*), already. To prove (6.21.*a*) we perturb $\lambda_0$, $\sigma_+^+(\lambda_0)$ to $\lambda_0\pm\delta$, $\sigma_+^+(\lambda_0\pm\delta)$. As $\lambda$ decreases from $\lambda_0$ to $\lambda_0-\delta$, i.e. along the stationary branch, the critical Floquet exponent satisfies

$$\eta \quad = \quad \alpha(\lambda)\,.$$

This Floquet exponent is (algebraically and geometrically) double because it is associated to the real eigenspace of $\alpha(\lambda)\pm i\beta(\lambda)$. Therefore

$$\sigma_+^+(\lambda_0-\delta) \quad \equiv \quad \sigma_+^+(\lambda_0)\,. \tag{6.23}$$

As $\lambda$ increases from $\lambda_0$ to $\lambda_0+\delta$, i.e. along the periodic branch, one of the two critical Floquet exponents is simple trivial, $\eta = 0$, with eigenfunction $\frac{d}{d\tau}\tilde{\xi}_\lambda$. The other critical exponent $\eta$ is also simple, but becomes nonzero and satisfies

$$\begin{aligned} sign\ \eta &= -sign\,(D_s^2\lambda(0)\cdot D_\lambda\alpha(\lambda_0)) = \\ &= -sign\ D_\lambda\alpha(\lambda_0) = -\chi^r \end{aligned} \tag{6.24}$$

by [Cra&Rab2], cf. also [Fie2, (1.3)]. This is the standard formula for exchange of stability at a local Hopf bifurcation. Remember that $D_s^2\lambda(0) > 0$, for our case $\lambda(s) > \lambda_0$. Now (6.24) implies

$$\sigma_+^+(\lambda_0+\delta) \equiv \begin{cases} \sigma_+^+(\lambda_0) & \text{if } \chi^r = +1, \\ \sigma_+^+(\lambda_0)+1 & \text{if } \chi^r = -1. \end{cases} \tag{6.25}$$

Together, (6.23) and (6.25) imply (6.21.$a$). This proves relation (6.20) between $\Phi$ and ⏀, if $\lambda(s) > \lambda_0$ for the bifurcating branch.

In case bifurcation occurs to the left, $\lambda(s) < \lambda_0$, just replace $\lambda$ by $-\lambda$. This reflection in $\lambda$ reverses the signs of $\lambda(s) - \lambda_0$, of $\chi^r$ and, consequently, of ⏀$^r(\lambda_0, x_0)$. But $\Phi$ remains unchanged. Hence (6.20) remains valid. This completes the proof of proposition 6.3. □

## 6.5 Proof of theorem 2.7 for finite cyclic groups

Building onto propositions 6.2 and 6.3, we can now prove theorem 2.7 on generic global Hopf bifurcation in the case of finite cyclic symmetry $G = \mathbb{Z}_n$. This case will also serve as a paradigm for $G = \mathbb{Z}_\infty$, later.

**Proof of theorem 2.7** ($n < \infty$) **:**

Recall that we assume $G = \mathbb{Z}_n, n < \infty$, and that the nonlinearity $f$ is generic $G$-equivariant so that theorem 5.2 on generic centers and theorem 5.11 on generic secondary bifurcation both hold. Moreover we assume

$$\mathcal{X}_n^d = \sum_{r\in d} ⏀^r \neq 0 \tag{2.17}$$

for some binary orbit $d \in D(n)$, cf. definitions 2.3-2.5.

Similarly to the proof in the case of no symmetry (cf. §3), we define equivariant snakes $S \subseteq \Lambda \times X$. Let $Z^* \subseteq \Lambda \times X$ be the set containing all $G$-centers $(\lambda_0, x_0)$ with representation $r \in d$, all periodic solutions $(\lambda, x(t))$ with symmetry $(G, K, \Theta)$ such that

$$\Theta \in d \quad \text{and} \quad \Phi \neq 0$$

for the orbit index $\Phi$ of $(\lambda, x(\cdot))$, and containing in addition the bifurcation orbits from table 5.2. Any maximal connected component of $Z^*$ which contains a center is called an **equivariant snake**. If there exists a **global snake** $S$, i.e. an equivariant snake which is unbounded in $\mathbb{R} \times X$ or contains periodic solutions with arbitrarily large minimal period, then we may put

$$Z := S \tag{6.26}$$

and theorem 2.7 is proved.

Otherwise, each snake $S$ is bounded together with its minimal periods. We claim, and prove below, that each snake $S$ then contains not only one but exactly two centers; one with

$$\text{\textphi}^{r} = +1,$$

called the **source** of $S$, and one **sink** with

$$\text{\textphi}^{r'} = -1,$$

$r, r' \in d$. In other words, sources and sinks are joined in pairs by snakes. This implies

$$\mathcal{X}_n^d = \sum_{r \in d} \text{\textphi}^{r} = 0 \tag{6.27}$$

because, conversely, each center with representation $r \in d$ lies on a snake, by proposition 6.3. But (6.27) contradicts our assumption (2.17) of nonzero $\mathcal{X}$-index. Therefore a global snake does exist, and theorem 2.7 holds by (6.26).

It remains to prove our claim that each bounded snake $S$ contains exactly one source and one sink. Indeed $S$ contains a center $(\lambda_0, x_0)$, by definition, which we may assume to be a source. Otherwise just replace $\lambda$ by $-\lambda$, reversing the sign of $\text{\textphi}$. Locally near any periodic trajectory of type 0, we may orient the curve $S$ of periodic trajectories such that

$$\Phi \cdot \lambda \quad \text{increases} \tag{6.28}$$

along $S$. Note that $\Phi$ is locally constant at such trajectories by homotopy invariance proposition 6.2. Near secondary bifurcations $(\lambda^*, x^*_\lambda)$, exactly two of the emanating branches belong to $S$, again by homotopy invariance proposition 6.2, and because $\Theta$ remains in $d$, cf. corollary 5.13. Moreover, the orientation (6.28) can be extended consistently through $(\lambda^*, x^*_\lambda)$. The snake $S$ cannot loop back onto a point on itself, by this construction. However, $S$ is bounded together with its minimal periods. Thus $S$ terminates at another center $(\lambda_0', x_0')$. The orientation (6.28) of the entire snake $S$ makes the source $(\lambda_0, x_0)$ the starting point of $S$, by proposition 6.3. Again by proposition 6.3, the end point $(\lambda_0', x_0')$ of the oriented snake $S$ then must be a sink, as was claimed above. This completes the proof of theorem 2.7 in case $G = \mathbb{Z}_n$ is a finite cyclic group. □

## 6.6 The case of the infinite cyclic group

We now return to the case of infinite cyclic $G = \mathbb{R}/\mathbb{Z} = \mathbb{Z}_\infty$. First we define an orbit index $\Phi$ for concentric and for rotating waves. In fact this definition coincides with the one given by Mallet-Paret & Yorke [M-P&Y2] in the no-symmetry case. Then we show that propositions 6.2 and 6.3 remain valid for this orbit index and for $G = \mathbb{Z}_\infty$. Finally we prove theorems 2.6 and 2.7 for that case.

To define the orbit index $\Phi$ for $G = \mathbb{Z}_\infty$ we recall the operator setting $F(f, \lambda, \xi, p) = 0$ for periodic solutions $x$ rescaled to $\xi$, cf. (6.2), (6.3). We also recall from (6.5) and (6.6) the spaces $\mathcal{E}^a$, $a \in \{+, -\}$, given by $y(\tau + 1) = ay(\tau)$. These spaces are invariant under $D_\xi F(f, \lambda, \xi, p)$. Define

$$\sigma^a : \quad \text{the number of positive eigenvalues } \eta \text{ of } D_\xi F \text{ on } \mathcal{E}^a, \text{ counting algebraic multiplicities.} \tag{6.29}$$

**6.4 Definition :**
*Let $G = \mathbb{Z}_\infty$ and let $x(t)$ be a periodic solution with symmetry $(G, K, \Theta)$. We define the* **orbit index $\Phi$** *of $x$ by*

$$\Phi(x) \quad := \quad \begin{cases} (-1)^{\sigma^+} & \text{if } \sigma^- \text{ is even,} \\ 0 & \text{otherwise,} \end{cases} \tag{6.30}$$

*where $\sigma^+, \sigma^-$ are defined in (6.29) above. In other words*

$$\Phi(x) \quad = \quad \frac{1}{2}\left((-1)^{\sigma^+} + (-1)^{\sigma^+ + \sigma^-}\right). \tag{6.30$'$}$$

Note that the orbit index $\Phi$ indeed coincides with the definition (3.11) given by Mallet-Paret & Yorke [M-P&Y2], by relations (6.4) and (6.7) between Floquet multipliers $\mu$ and Floquet exponents $\eta$. Obviously, $\sigma^+$ counts real multipliers $\mu > 1$ while $\sigma^-$ counts multipliers $\mu < -1$. In the case of concentric waves, $G = K = \mathbb{Z}_\infty$, the orbit index $\Phi$ also coincides with the "no-symmetry" orbit index which we obtain when working in the subspace $X^K$ instead of $X$. Indeed the multiplicities of any eigenvalue $\eta$ of $D_\xi F$ on $\mathcal{E}^a \subseteq \tilde{C}^0(\mathbf{R}, X)$ and on $\mathcal{E}^a \cap \tilde{C}^0(\mathbf{R}, X^K)$ coincide, *mod* 2. Just observe that $\tilde{C}^0(\mathbf{R}, X^K)$ is the representation subspace of the trivial representation of the subgroup $K \times \{0\} \cong \mathbb{Z}_\infty$ of $G \times S^1$, acting on $\tilde{C}^0(\mathbf{R}, X)$ as usual. All other representations occur with even real dimension in any eigenspace. With this in mind, it is now fairly easy to prove propositions 6.2, 6.3 and theorems 2.6, 2.7 for $G = \mathbb{Z}_\infty$.

**Proof of proposition 6.2 ($G = \mathbb{Z}_\infty$) :**
We consider type 0, turns, and doublings first. Because $\Phi$ coincides with the orbit index (3.11) of Mallet-Paret & Yorke by the remarks above, homotopy invariance of $\Phi$ follows from [M-P&Y2, Chow&M-P&Y2].

Alternatively we may invoke proposition 6.2 for finite $G$ putting $G := \{0\}$ artificially. That is, we ignore all group structure which is actually present. In this perspective the doubling bifurcations for concentric as well as for rotating waves in table 5.2 become flip doublings with $G = H = K = \{0\}$, $m = 1$. Any way, homotopy invariance of $\Phi$ is seen to hold for type 0, turns and doublings.

It remains to prove homotopy invariance of $\Phi$ at a freezing $(\lambda_0, x_0)$, cf. table 5.2 (b.3). From §5.3 we recall the setting (5.2-5.7) and in particular definition 5.3 of a freezing. We denote the symmetry of the freezing $(\lambda_0, x_0)$ by $(G, K, \pm\Theta)$. We also recall the following expression for the linearized flow $\Psi_t$ for the local branch $(\lambda, x_\lambda)$ of rotating waves

$$D_x\Psi_p(f, \lambda, x_\lambda) \quad = \quad \rho(1/\Theta)\, exp\Big(p\,(-\alpha\mathcal{R} + D_x f(\lambda, x_\lambda))\Big). \tag{5.5.b}$$

Here $\mathcal{R}$ denotes infinitesimal rotation. Let $X^+$ resp. $X^-$ denote the representation subspaces of the + resp. − representation of the cyclic group $K = \{0, 1/\Theta, \ldots, (\Theta - 1)/\Theta\}$ on $X$, i.e. $\rho(1/\Theta) = \pm id$ on $X^\pm$. We transcribe $\sigma^a$ from definition (6.29), (6.30) of the orbit index $\Phi$ into Floquet-multipliers $\mu$, i.e. into eigenvalues of $D_x\Psi_p$. For $a, b \in \{+, -\}$, $\lambda \neq \lambda_0$ let

$$\begin{aligned} \hat\sigma^a_b(\lambda): \quad & \text{the number of real eigenvalues } \mu \text{ of} \\ & \qquad D_x\Psi_p(f, \lambda, x_\lambda)|_{X^b} \;:\; X^b \to X^b \\ & \text{with } a\mu > 1, \text{ counting algebraic multiplicity.} \end{aligned} \tag{6.31}$$

Now recall that $K$ commutes with $D_x\Psi_p$. Moreover, all other representation spaces of $K$, except the +,− representation spaces, have even dimension in any eigenspace of $D_x\Psi_p$. Hence

$$\sigma^a(\lambda) \equiv \hat{\sigma}^a_+(\lambda) + \hat{\sigma}^a_-(\lambda) \qquad (mod\ 2)$$

Therefore homotopy invariance of $\Phi$ across $(\lambda_0, x_0)$ follows, if we can show for any $a, b \in \{+,-\}$ and for any small positive $\delta$ that

$$\hat{\sigma}^a_b(\lambda_0 + \delta) \equiv \hat{\sigma}^a_b(\lambda_0 - \delta) \qquad (mod\ 2)\,. \tag{6.32}$$

To prove (6.32), we recall from (5.5.b) and the definition of $X^b$ that on $X^b$

$$D_x\Psi_p = b \cdot exp\Big(p(-\alpha\mathcal{R} + D_x f(\lambda, x_\lambda))\Big)\,.$$

Thus the $\hat{\sigma}^a_b(\lambda)$ remain unchanged (*mod* 2) if we replace $D_x\Psi_p$ on $X^b$ by

$$b \cdot exp\Big(-\alpha\mathcal{R} + D_x f(\lambda, x_\lambda)\Big)\,.$$

For $\lambda = \lambda_0$, i.e. for $\alpha = 0$, the only eigenvalue on the unit circle of this operator on $X^+$ is the trivial simple eigenvalue $b = +1$ with eigenvector $\mathcal{R}x_\lambda$, because the spectrum of $D_x f(\lambda_0, x_0)$ on the imaginary axis consists of only a simple eigenvalue zero, cf. (5.7.a). Similarly, there is no eigenvalue on the unit circle when we consider $X^-$. This situation persists for $\lambda$ near $\lambda_0$, by a perturbation argument. Therefore (6.32) holds, and the proof of proposition 6.2 is complete. □

**Proof of proposition 6.3** ($G = \mathbb{Z}_\infty$) **:**

For rotating waves, $\Phi$ and $\Phi^r$ both coincide with the orbit and center index of Mallet-Paret & Yorke [M-P&Y2], evaluated in $X$. Thus proposition 6.3 follows from [M-P&Y2], see also [Fie2, lemma 4.3]. Replacing $X$ by $X^K$ the previous two sentences also hold for concentric waves.

Alternatively, we could resort to the proof of proposition 6.3 with $G = \{0\}$, again, as in the proof of proposition 6.2 ($G = \mathbb{Z}_\infty$).

Either way, proposition 6.3 is proved for $G = \mathbb{Z}_\infty$. □

**Proof of theorem 2.7** ($G = \mathbb{Z}_\infty$) **:**

With propositions 6.2 and 6.3 being proved for $G = \mathbb{Z}_\infty$, too, the proof of theorem 2.7 given in §6.5 for finite $G = \mathbb{Z}_n$ applies verbatim, replacing $n < \infty$ by $n = \infty$. This proves theorem 2.7 for $G = \mathbb{Z}_\infty$. □

**Proof of theorem 2.6 :**

Again, we employ the proof of theorem 2.7 for finite $G = \mathbb{Z}_n$, replacing $n < \infty$ by $n = \infty$. This time, however, we call a snake $S$ global only if it is unbounded in $\mathbb{R} \times X$. If a global snake $S$ exists we may put $Z := S$ and theorem 2.6 is proved.

Otherwise, each snake $S$ is bounded. But the minimal period on $S$ might be unbounded. From table 5.2(b) we recall that for rotating waves this can occur only at a freezing. Note that $\Phi$ does not change at a freezing, by proposition 6.2. However, $\Theta$ does change sign. Following the oriented bounded snake $S$ from source to sink, as in the proof of theorem 2.7, we can therefore only conclude that

$$\mathcal{H}_{\infty}^{d} + \mathcal{H}_{\infty}^{-d} = \sum_{r \in d \cup (-d)} \Phi^{r} = 0. \qquad (6.27)'$$

Still, this is a contradiction to our assumption

$$\mathcal{H}_{\infty}^{d} + \mathcal{H}_{\infty}^{-d} \neq 0. \qquad (2.15)$$

This proves theorem 2.6. □

# 7 General global theory

## 7.1 Outline

The generic theory which we have developed so far suffers from its characteristic but very restrictive genericity assumption (2.13). See theorems 2.6, 2.7 and their proofs in the previous section. Proving theorems 2.9, 2.10 we dispose of this awful genericity assumption altogether. However, the notions of virtual symmetry and of virtual period as presented in definition 1.2 and in §4 now gain their full weight.

After a little preparation on elementary set topology we first prove theorem 2.10 and then theorem 2.9 by generic approximation. Afterwards we prove corollary 2.13 which states that "everything" also works for analytic semigroups. We use this proof to briefly summarize the long line of theoretical reasoning which we have pursued so far, and which comes to its end in this chapter.

## 7.2 Convergence of continua

To motivate our excursion into set topology let us recall our basic strategy for a proof of theorems 2.9 and 2.10, as outlined in §1.5 and §3.5. We are now given a general, not necessarily generic nonlinearity $f \in \mathcal{F}$ (see (1.36)) satisfying assumptions (2.20.a-c). We approximate $f$ by generic nonlinearities $g_i \in \mathcal{F}$, for which theorems 2.6, 2.7 apply. This gives us global Hopf bifurcation for $g_i$ in form of a continuum $Z_i$. Naively, we would like to pass to a "limiting continuum"

$$\mathcal{C} \quad = \quad \text{"lim"}\ Z_i$$

which should provide global Hopf bifurcation for $f$ itself. Because life is not all that simple, we are forced to some meandering.

Let us define a limit for any sequence $S_i$ of subsets of some metric space. Following Whyburn [Why], we define

$$\begin{aligned} \limsup S_i \quad &:= \quad \{z \mid z = \lim z_{i_j} \text{ for some sequence } z_{i_j} \in S_{i_j},\ i_j \to \infty\} \qquad (7.1) \\ &= \quad \bigcap_{i_0 \geq 1} clos \left( \bigcup_{i \geq i_0} S_i \right), \end{aligned}$$

where *clos* denotes closure.

**7.1 Lemma** [Why] :

*Assume that*

$$\bigcup S_i \text{ is relatively compact,} \qquad (7.2.a)$$

$$\text{each } S_i \text{ is connected, and} \qquad (7.2.b)$$

$$\text{there exists a converging sequence } (z_i)_{i=1}^{\infty} \text{ with } z_i \in S_i\,. \qquad (7.2.c)$$

*Then* $\limsup S_i$ *is nonempty, compact, and connected.*

**Proof :**

Obviously $\limsup S_i$ is nonempty because it contains $\lim z_i$. By [Why, §I.9] $\limsup S_i$ is connected. Being an intersection of closed sets, $\limsup S_i$ is closed. By assumption (7.2.a) it is therefore also compact. This completes the proof. □

## 7.3 Proof of theorem 2.10

We are given a cyclic factor $H_0/K_0 \cong \mathbb{Z}_n$ of the group $\Gamma$, and a binary orbit $d \in D(n)$ such that

$$\mathcal{X}^{\pm d}_{H_0,K_0} \neq 0.$$

To construct a global continuum $\mathcal{C} \subseteq \Lambda \times X^{K_0}$ of $H_0$-centers and of periodic solutions of $f$, we first construct a sequence $\mathcal{C}_k$ of ever larger continua for $f$ with the desired virtual symmetries. Each $\mathcal{C}_k$ will be bounded. But all $\mathcal{C}_k$ will have a center

$$z_0 = (\lambda_0, x_0)$$

of $f$ in common. For $k$ sufficiently large, $\mathcal{C}_k$ will not be contained in the set of centers of $f$. Rather, $\mathcal{C}_k$ will contain some periodic solution with large norm in $\Lambda \times X^{K_0}$, or with large virtual period. Putting

$$\mathcal{C} := \bigcup_k \mathcal{C}_k$$

will then finish the proof.

Let us first fix a large constant $c_0 > 0$ throughout the proof such that the following conditions (7.3.a-c) all hold.

The set of centers of $f$ in $\Lambda \times X^{K_0}$ is contained in the open ball $\mathcal{B}_{c_0} \subseteq \Lambda \times X^{K_0}$ around (0,0) with radius $c_0$. (7.3.a)

Any purely imaginary eigenvalue $\pm i\beta$ at any center of $f$ in $\Lambda \times X^{K_0}$ satisfies (7.3.b)

$$2\pi/|\beta| < c_0.$$

Lemma 4.11 holds for $c_0$. That is, the set $VP(c_0)$ of those centers for which all virtual periods are less than $c_0$ is dense in the set of all centers. (7.3.c)

Note that (7.3.a, b) can be satisfied by assumptions (2.20.a, b) on $f$. For (7.3.c) we need the analyticity assumption (2.20.c) which enters into lemma 4.11.

To construct $\mathcal{C}_k$, $k \geq c_0$, we choose a generic approximation

$$f = \lim g_i. \tag{7.4}$$

More precisely, the $g_i \in \mathcal{F}$ are assumed to be generic in the sense of definition 1.3 so that theorems 2.7, 5.2, and 5.11 hold for the action of the cyclic group

$$G := H_0/K_0 \cong \mathbb{Z}_n, \qquad n \leq \infty$$

on $X^{K_0}$. Convergence in (7.4) is understood as uniform $C^1$-convergence on bounded subsets of $\Lambda \times X^{K_0}$. Moreover, we may assume that assumptions (2.20.$a,b$) and likewise (7.3.$a,b$) still hold if we replace $f$ there by $g_i$, for any $i$. Indeed, all requirements on $g_i$ concerning centers outside of $\mathcal{B}_{c_0}$ are satisfied automatically if no such centers arise by the perturbation $g_i$ of $f$.

Denoting the global equivariant Hopf index for $f$ resp. $g_i$ by $\mathcal{H}^{\pm d}_{H_0,K_0}(f)$ resp. $\mathcal{H}^{\pm d}_{H_0,K_0}(g_i)$, this implies that for all $i \geq i_0$ (say, for all $i$)

$$\mathcal{H}^{\pm d}_{H_0,K_0}(g_i) \quad = \quad \mathcal{H}^{\pm d}_{H_0,K_0}(f) \quad \neq \quad 0\,. \tag{7.5}$$

Indeed, $g_i \to f$ uniformly in $C^1$ on the bounded set $\mathcal{B}_{c_0}$. With the above properties of $g_i$, (7.5) follows from definition 2.8 of $\mathcal{H}^{\pm d}_{H_0,K_0}$. By (2.27), $\mathcal{H}^{\pm d}_{H_0,K_0}(g_i)$ relates to the generic equivariant Hopf index $\mathcal{H}^d_n(g_i)$:

$$\mathcal{H}^{\pm d}_{H_0,K_0}(g_i) \quad = \quad \mathcal{H}^d_n(g_i) + \mathcal{H}^{-d}_n(g_i)\,.$$

Without loss of generality, we thus may assume that, for all $i$,

$$\mathcal{H}^d_n(g_i) \quad \neq \quad 0\,. \tag{7.5)$'$ (}$$

Therefore theorem 2.7 provides us, for each $g_i$, with a global continuum $Z_i \subseteq \Lambda \times X^{K_0}$ of generic $H_0$-centers with representations $r \in d$ of $H_0/K_0 \cong G$, and of periodic solutions with symmetry $(H_0, K, \Theta)$, $K \geq K_0$, $\Theta \in d$. Cutting off $Z_i$ suitably beyond a ball $\mathcal{B}_k$ or for minimal periods $\geq 2k$ we may elaborate on this continuum as follows. For any $i$ and any $k \geq c_0$, we claim that there exist connected compact sets ("snakes") $S_{i,k} \subseteq clos(\mathcal{B}_k)$ with the following properties (7.6.$a$-$d$):

$S_{i,k}$ contains exactly one center (7.6.$a$)

$$(\lambda_i, x_i) \quad \in \quad \mathcal{B}_{c_0}$$

of $g_i$, and periodic solutions with symmetry $(H_0, K, \Theta)$, $\Theta \in d$;

the minimal periods $p$ of the periodic solution vary continuously on $S_{i,k}$, except for possible jumps by a factor 2; (7.6.$b$)

$$p \leq 2k \quad \text{on} \quad S_{i,k}\,; \tag{7.6.c}$$

one of the following two conditions holds: (7.6.$d$)

(i) $S_{i,k}$ hits the boundary of $\mathcal{B}_k$ or

(ii) there exists a nonempty, compact, connected subset $S'_{i,k} \subseteq S_{i,k}$ and elements $(\underline{\lambda}_{i,k}, \underline{x}_{i,k}), (\bar{\lambda}_{i,k}, \bar{x}_{i,k})$ with minimal periods $\underline{p}, \bar{p}$, such that

$$c_0 \leq \underline{p} \leq 2c_0\,, \quad k \leq \bar{p} \leq 2k$$

and all minimal periods $p$ on $S'_{i,k}$ satisfy

$$c_0 \leq p \leq 2k\,.$$

Indeed, the $S_{i,k}$ are subsets of the global snake $Z_i$ which bifurcates from the center $(\lambda_i, x_i)$. By the generic bifurcation diagrams of theorem 5.11, minimal periods can only jump by factors 2. Moreover $Z_i$ is a global continuum and, by (7.3.$b$), $Z_i$ does contain minimal periods $\underline{p} \in [c_0, 2c_0]$. Thus (7.6.$a$-$d$) can be satisfied.

We fix $k \geq c_0$ and construct $C_k$. Passing to a subsequence of $i$, if necessary, we may assume that

$$\lim (\lambda_i, x_i) = (\lambda_0, x_0)$$

exists. Note that $(\lambda_0, x_0)$ is a center of $f$. As in (7.1) we define

$$C_k := \limsup_{i\to\infty} S_{i,k}. \tag{7.7}$$

Note that all elements of $C_k$ have the virtual symmetry properties desired for $C$. Indeed, this follows from virtual symmetry corollary 4.6 together with $S_{i,k} \subseteq clos\mathcal{B}_k$ and boundedness condition (7.6.$c$).

We claim that $C_k$ is compact and connected. We apply lemma 7.1. Assumptions (7.2.$b, c$) hold by construction. Trivially $\bigcup_i S_{i,k} \subseteq clos\mathcal{B}_k$, being bounded, is relatively compact. Thus $C_k$ is compact and connected.

Next we claim that $C_k$ is not contained in the set of centers of $f$, for $k \geq k_0 \geq c_0$ large enough. We have to consider two cases.

**Case 1** (7.6.$d$) – (i) is satisfied for a sequence $i_j \to \infty$ (which may depend on $k$).

Then $C_k$ hits the boundary of $\mathcal{B}_k$, but this boundary does not contain any centers of $f$, by (7.3.$a$) and $k \geq c_0$.

**Case 2** (7.6.d) – (ii) holds for all $i \geq i_0(k)$.

We give an indirect argument: suppose $C_k$ consists only of centers. Discussing the virtual periods on $C_k$ we will obtain a contradiction. Replacing $i$ by a subsequence, if necessary, we may assume that $(\bar{\lambda}_{i,k}, \bar{x}_{i,k})$ converge for $i \to \infty$. By lemma 7.1 again,

$$\emptyset \neq C'_k := \limsup_{i\to\infty} S'_{i,k} \subset C_k$$

is connected and consists of centers of $f$, only. Because all centers lie on one-dimensional branches, by assumptions (2.20.$a, b$), the set $C'_k$ can be viewed as an interval. Now note that

$$C'_k \cap VP(c_0) = \emptyset$$

because any element of $C'_k$ has some virtual period $\geq c_0$ by construction of $C'_k$ and by corollary 4.6 on virtual periods. The set $VP(c_0)$ was defined in (4.18). Because $VP(c_0)$ is dense in the set of centers, by lemma 4.11, the interval $C'_k$ therefore consists of only one single center $(\lambda'_0, x'_0)$. Via lemma 4.11 our analyticity assumption (2.20.$c$) enters here. By (7.6.$b$) and (7.6.$d$) – ($ii$), any interval $[t, 2t] \subseteq [c_0, 2k]$ contains at least one virtual period of $(\lambda'_0, x'_0)$. Thus $(\lambda'_0, x'_0)$ has at least $2^{j_0}$ different virtual periods provided that

$$2^{2^{j_0}} c_0 < 2k_0 \leq 2k.$$

This contradicts lemma 4.8 as follows. Let $j_0$ be a uniform upper bound for the number of different pairs of purely imaginary eigenvalues at centers of $f$. By lemma 4.8, $(\lambda'_0, x'_0)$ then has less than $2^{j_0}$ distinct virtual periods, contradicting the above. Therefore, $C'_k \subseteq C_k$ does not consist only of centers. This closes case 2.

Finally, we define the continuum

$$C := \bigcup_{k\geq k_0} C_k.$$

Note that $C$ is connected since the $C_k$ are and since $(\lambda_0, x_0) \in C_k$, for all $k$. Following the reasoning in cases 1,2 above, we see that for any $k \geq k_0$ there exists a nonstationary periodic solution $(\lambda, x) \in C$ such that $|(\lambda, x)| = k$, or such that a virtual period of $(\lambda, x)$ is $\geq 2^{-2^{j_0}} k$. Thus $C$ is indeed global. Recalling that all solutions in $C$ have the desired virtual symmetries completes the proof of theorem 2.10. □

**7.2 Remark :**

We observe that analyticity assumption (2.20.c) can be dropped in theorem 2.10 if the set of $H_0$-centers in $X^{K_0}$ is known to be discrete. Obviously, the above proof goes through in that case.

## 7.4 Proof of theorem 2.9

With proper modifications the proof follows the same scheme as the preceding proof of theorem 2.10. We construct a sequence of ever larger bounded continua $C_k$ with a common center $(\lambda_0, x_0)$ and define

$$C \quad := \quad \bigcup_k C_k \, ,$$

again. The main difference lies in the treatment of (virtual) period. Recall from §5.3 and in particular from (5.7) that $x$ is on a rotating resp. frozen wave iff

$$-\alpha \mathcal{R}x + f(\lambda, x) \quad = \quad 0, \qquad \mathcal{R}x \neq 0, \tag{7.8}$$

for some $\alpha \neq 0$ resp. $\alpha = 0$. Here $\mathcal{R}$ is the infinitesimal generator of $H_0/K_0 \cong \mathbb{Z}_\infty$ on $X^{K_0}$. Also recall that for a rotating wave

$$\alpha \quad = \quad (\Theta p)^{-1},$$

where $p$ is the minimal period and $(H_0, K, \Theta)$, $K \geq K_0$, is the symmetry, as usual. The analogous statements hold with $f$ replaced by $g_i$, of course.

By assumption (2.20.b), again, we may fix a large constant $c_0 > 0$ such that

the set of centers of $f$ in $\Lambda \times X^{K_0}$ is contained in the open ball $\mathcal{B}_{c_0} \subseteq \Lambda \times X^{K_0}$ around (0,0) with radius $c_0$. (7.9)

Further we choose an approximation

$$f \quad = \quad \lim g_i$$

by generic $g_i$, as in the proof of theorem 2.10, such that (2.20.a, b) hold for $g_i$ as well. By uniform $C^1$-convergence on bounded sets, this implies

$$\mathcal{X}^{\pm d}_{H_0,K_0}(g_i) \quad = \quad \mathcal{X}^{\pm d}_{H_0,K_0}(f) \quad \neq \quad 0$$

for $i \geq i_0$ (say for all $i$), and hence

$$\mathcal{X}^{d}_{\infty}(g_i) \; + \; \mathcal{X}^{-d}_{\infty}(g_i) \quad \neq \quad 0$$

without loss of generality.

Cutting off the global continuum $Z_i$ which theorem 2.6 provides for any $g_i$, we find a compact connected set ("snake") $S_{i,k} \subseteq clos(\mathcal{B}_k)$, for each $i$ and $k \geq k_0 \geq c_0$ large enough, with the following properties (7.10.a-c):

$S_{i,k}$ contains exactly one center (7.10.$a$)

$$(\lambda_i, x_i) \quad \in \quad \mathcal{B}_{c0}$$

of $g_i$, and rotating or frozen waves with symmetry $(H_0, K, \Theta)$ resp. $(H_0, K, \pm\Theta)$, $K \geq K_0$, $\Theta \in (d \cup (-d))$;

$|\alpha| \leq k$ and $|\Theta| \leq 2k$ on $S_{i,k}$ for $\alpha$ as in (7.8); (7.10.$b$)

one of the following two conditions holds: (7.10.$c$)
(i) $S_{i,k}$ hits the boundary of $\mathcal{B}_k$, or
(ii) $S_{i,k}$ contains a rotating or a frozen wave $(\bar\lambda_{i,k}, \bar x_{i,k})$ with symmetry $(H_0, K_{i,k}, \Theta_{i,k})$ resp. $(H_0, K_{i,k}, \pm\Theta_{i,k})$, $K_{i,k} \geq K_0$, such that $|\Theta_{i,k}| \in \{k+1, k+2, \cdots, 2k\} \cap (d \cup (-d))$.

Indeed, $S_{i,k}$ are subsets of the global snake $Z_i$ which bifurcates from the center $(\lambda_i, x_i)$. We choose $i, k$ large enough so that (7.10.$a$) and (7.10.$b$) do not contradict each other. For (7.10.$c$) we use that $Z_i$ is unbounded. Note that $\{k+1, \cdots, 2k\} \cap (d \cup (-d))$ consists of exactly one element. Hence $|\Theta_{i,k}|$ is independent of $i$ and, passing to a subsequence, we may assume

$$\Theta_{i,k} \quad =: \quad \bar\Theta_k \quad > \quad 0\,.$$

Likewise $K_{i,k} =: \bar K_k$ is independent of $i$. We need the bounds (7.10.$b$) to construct limits $C_k$, applying virtual symmetry corollary 4.6. We need (7.10.$c$) to make $C = \bigcup C_k$ global. Conditions (7.10.$a$-$c$) are analogous to conditions (7.6.$a$-$d$) from the proof of theorem 2.10.

Assuming again that

$$\lim\, (\lambda_i, x_i) \quad = \quad (\lambda_0, x_0)$$

exists we obtain compact connected sets

$$C_k \quad := \quad \lim_{i\to\infty} S_{i,k}\,,$$

as before. By the second part of virtual symmetry corollary 4.6, the continuum $C_k$ consists of rotating waves, of frozen waves, and of centers, with the desired virtual symmetry properties. This uses $S_{i,k} \subseteq clos\, \mathcal{B}_k$ and the bounds (7.10.$b$). Note that case (4.9.$d$) cannot occur because stationary solutions of $f$ in $X^{H_0}$ are nondegenerate if linearized in $X^{K_0}$, by assumption (2.20.$a$).

We show indirectly that

$$C \quad := \quad \bigcup_{k \geq k_0} C_k$$

is unbounded in $\Lambda \times X^{K_0}$. Suppose $C$ is bounded. Then $S_{i,k}$ satisfies case ($ii$) of (7.10.$c$) for all but finitely many $i$, provided that $k$ is fixed large enough. Thus there exist $(\bar\lambda_{i,k}, \bar x_{i,k})$ satisfying

$$-\alpha_{i,k} \mathcal{R} \bar x_{i,k} + g_i(\bar\lambda_{i,k}, \bar x_{i,k}) \quad = \quad 0$$

with $\alpha_{i,k} = 0$ (frozen wave), or with $\alpha_{i,k} = (p_{i,k} \bar\Theta_k)^{-1}$ (rotating wave). Passing to a subsequence, if necessary, we may assume that

$$(\alpha_k, \bar\lambda_k, \bar x_k) \quad := \quad \lim_{i\to\infty} (\alpha_{i,k}, \bar\lambda_{i,k}, \bar x_{i,k})$$

exists, satisfying

$$-\alpha_k \mathcal{R}\bar{x}_k + f(\bar{\lambda}_k, \bar{x}_k) = 0. \tag{7.11}$$

Because $(\bar{\lambda}_k, \bar{x}_k) \in C_k$ and $C = \bigcup C_k$ is assumed bounded, we may even assume that

$$(\bar{\lambda}, \bar{x}) := \lim_{k \to \infty} (\bar{\lambda}_k, \bar{x}_k)$$

exists. Moreover

$$\lim_{k \to \infty} \alpha_k = 0,$$

even in case $\alpha_{i,k} = (p_{i,k}\bar{\Theta}_k)^{-1}$. Indeed, $\bar{\Theta}_k \geq k$, and there exists a uniform lower bound on the minimal periods $p_{i,k}$ on bounded subsets of $\Lambda \times X$, which depends only on a uniform bound for the Lipschitz constants of the $g_i$. This implies

$$f(\bar{\lambda}, \bar{x}) = 0.$$

Now suppose that

$$\mathcal{R}\bar{x}_k \neq 0 \qquad \text{for all } k \text{ large enough.} \tag{7.12}$$

Multiplying (7.11) by $|\mathcal{R}\bar{x}_k|^{-1}\mathcal{R}$ and denoting $|\mathcal{R}\bar{x}_k|^{-1}\mathcal{R}\bar{x}_k$ by $\bar{y}_k$ we obtain from (7.11) the linearization

$$-\alpha_k \mathcal{R}\bar{y}_k + D_x f(\bar{\lambda}_k, \bar{x}_k)\bar{y}_k = 0. \tag{7.13}$$

Choosing a convergent subsequence

$$\bar{y} := \lim_k \bar{y}_k \in X^{K_0} \setminus \{0\}$$

we conclude that

$$D_x f(\bar{\lambda}, \bar{x})\,\bar{y} = 0,$$

because $\lim \alpha_k = 0$. Because $(\bar{\lambda}, \bar{x})$ is a stationary solution of $f$, this contradicts our nondegeneracy assumption (2.20.$a$).

Thus $C$ is unbounded provided that(7.12) holds. Suppose next that (7.12) does not hold, i.e.

$$\mathcal{R}\bar{x}_k = 0 \qquad \text{for arbitrarily large } k.$$

Then $\bar{x}_k \in X^{H_0}$ is not a rotating or frozen wave but, by virtual symmetry corollary 4.6, $(\bar{\lambda}_k, \bar{x}_k)$ is a center with some virtual symmetry at least $(H_0, \bar{K}_k, \bar{\Theta}_k)$. As usual, this implies that some virtual symmetry $(\hat{H}_k, \hat{K}_k, \hat{\Theta}_k)$ of $(\bar{\lambda}_k, \bar{x}_k)$ satisfies

$$\hat{K}_k \cap H_0 = \bar{K}_k.$$

Now lemma 4.8 on virtual symmetries of centers implies that

$$\bar{K}_k = \hat{K}_k \cap H_0 \leq K_j \cap H_0,$$

where $K_j = \Gamma_{(x_k, y_j)}$ for some eigenvector $y_j$ as in lemma 4.8. On the other hand

$$|\bar{K}_k| = |\bar{\Theta}_k| \geq k. \tag{7.14}$$

Because the set of all $H_0$-centers is compact, by (2.20.$b$), and because the total multiplicity of purely imaginary eigenvalues is finite at any fixed center, (7.14) is impossible if we choose $k$ large enough.

By this final contradiction, $C$ is indeed unbounded and the proof is complete. □

## 7.5 Proof of corollary 2.13

We briefly summarize the modifications which are necessary to prove theorems 2.6, 2.7, 2.9, 2.10 for an analytic semigroup setting under the additional assumptions (2.32.*a*-*d*). Mainly we use that compactness of the resolvent induces compactness of the semiflow.

In the proof of virtual symmetry corollary 4.6, we now construct the Fredholm operators $F_i$ slightly differently from (4.1):

$$F_i \quad := \quad -\xi + \left(\frac{1}{p_i}\frac{d}{dt} - A(\lambda_i)\right)^{-1} f_i(\lambda_i, \xi) \ .$$

As spaces we take $\Xi = \Xi' := \tilde{C}^0(\boldsymbol{R}, X)$. Note that we may assume the resolvent

$$\left(\frac{1}{p_i}\frac{d}{dt} - A(\lambda_i)\right)^{-1}$$

to exist. The resolvent is compact, hence $F_i$ is indeed Fredholm.

The local bifurcation theory of §5 remains unchanged in the semigroup setting. For a proof of genericity (§10) in the case of no symmetry see [Fie2]. The details for the symmetry case can be completed along those lines, following §10. In particular, the transversality theorem 10.2 is already adapted to this treatment.

Using Floquet theory in the semigroup setting, our construction of the equivariant orbit index $\Phi$, its homotopy invariance, and its relation to the center index ф given in §6 remain unchanged: these are all local considerations based on the results of §5. Because this establishes the same generic global pictures as in the finite dimensional case, the generic global theorems 2.6 and 2.7 can be proved as in §6.

The first modification in §7 concerns the proof of the compactness assumption (7.2.*a*) of lemma 7.1 on the sets $S_i := S_{i,k}$. Assumption (7.2.*a*) holds because, for $g_i \to f$, the set of points on periodic or stationary solutions in some fixed ball $B_k$ is relatively compact, again by compactness of the semiflow. The second modification concerns the limit $\bar{y} = \lim \bar{y}_k$ in the proof of theorem 2.9, cf. (7.13). In the semigroup case we obtain

$$-\alpha_k \mathcal{R} \bar{y}_k + A(\bar{\lambda}_k)\bar{y}_k + D_x f(\bar{\lambda}_k, \bar{x}_k)\bar{y}_k = 0, \tag{7.13$'$}$$

instead of (7.13). Equivalently,

$$-\bar{y}_k = A(\bar{\lambda}_k)^{-1}(-\alpha_k \mathcal{R}\bar{y}_k + D_x f(\bar{\lambda}_k, \bar{x}_k)\bar{y}_k)$$

is a relatively compact sequence by compactness of the resolvent $A(\bar{\lambda}_k)^{-1}$. Thus we may pass to the limit $\bar{y}$ and conclude

$$A(\bar{\lambda})\bar{y} + D_x f(\bar{\lambda}, \bar{x})\bar{y} = 0$$

at the stationary solution $(\bar{\lambda}, \bar{x})$:

$$A(\bar{\lambda})\bar{x} + f(\bar{\lambda}, \bar{x}) = 0 .$$

This contradicts the nondegeneracy of $(\bar{\lambda}, \bar{x})$ as in the earlier proof of theorem 2.9. With these remarks, corollary 2.13 is proved. □

# 8 Applications

## 8.1 Outline

We illustrate our main theorems 2.9 and 2.10 for rotating and discrete waves with a prototypical example: reaction diffusion systems with Brusselator kinetics, cf. [Nic&Pri]. In §8.2 we consider a ring of $n$ identical well-stirred Brusselator cells which are coupled symmetrically by diffusion along the sides of an $n$-gon. Such "Turing rings" [Tu] have attracted considerable interest from both the experimental and the mathematical point of view, see e.g. [Ale3, Ale4, Ale&Au2, Ale&Fie, Aro&Doe&Oth, Ash&Oth, Cro&Field, Go&St3, How, Oth, Oth&Scr, Schr&Hol&Kub&Marek, Schr&Marek1,2, Sma3, Stu&Marek, Swi]. Below we use the $D_n$ (dihedral) symmetry of Turing rings to find global continua of discrete waves. Local bifurcation is studied extensively in [Go&St3, Go&Sch&St, Swi]. We compare our global results to those of Alexander & Auchmuty [Ale&Au2]. As a second example, we consider the Brusselator system with diffusion in a three-dimensional ball, i.e. under $O(3)$ symmetry, see §8.3. We find global bifurcation of rotating waves, and of discrete waves with tetrahedral symmetry. Local bifurcation with such symmetries, among others, is analysed thoroughly in [Go&Sch&St, Go&St1]. To our knowledge, tetrahedral waves have not yet been observed experimentally.

Admittedly the Brusselator is not the most realistic model for the Belousov-Zhabotinskii reaction, and one might prefer other models (Oregonator, Field-Körös-Noyes model etc.) before comparing theoretical results with experiments, see e.g. [Field, Ty1, Ty2]. We choose the Brusselator as the simplest prototype, expecting similar waves but possibly different stability properties for other models. We conclude this section in §8.4 with further examples (hypercycle, graphs of oscillators, heterogeneous catalysis, fluid dynamics, etc.) which are amenable to our theory. We only sketch these applications phenomenologically, omitting any technical details.

## 8.2 Coupled oscillators

The $n$-gon of diffusively coupled Brusselators is given by the following differential equations:

$$\begin{aligned} \dot{x}_{i,1} &= \alpha^2(x_{i+1,1} - 2x_{i,1} + x_{i-1,1}) + f_1(x_{i,1}, x_{i,2}) \\ \lambda \dot{x}_{i,2} &= \alpha^2(x_{i+1,2} - 2x_{i,2} + x_{i-1,2}) + f_2(x_{i,1}, x_{i,2}) \end{aligned} \qquad (8.1\text{-}i)$$

for $i$ $(mod\ n)$, or in more condensed form

$$\begin{pmatrix} 1 & 0 \\ 0 & \lambda \end{pmatrix} \dot{x}_i = \alpha^2(\tilde{\Delta} x)_i + \tilde{f}(x_i). \qquad (8.1)$$

Here $x_i = (x_{i,1}, x_{i,2}) \in \mathbf{R}^+ \times \mathbf{R}^+$ is the concentration vector in the $i$-th cell, $x = (x_1, \ldots, x_n) \in \mathbf{R}^{2n}$,

$$(\tilde{\Delta} x)_i := x_{i+1} - 2x_i + x_{i-1}$$

is the "discretized Laplacian" describing diffusive coupling between neighboring cells, and $\alpha^2$ is a positive coupling constant. The Brusselator kinetics

$$\tilde{f}(u_1, u_2) = (f_1, f_2)(u_1, u_2)$$

is given by

$$\begin{aligned} f_1(u_1,u_2) &= 1-(b+1)u_1+u_1^2u_2 \\ f_2(u_1,u_2) &= bu_1-u_1^2u_2 \quad . \end{aligned} \tag{8.2}$$

Finally, $\lambda > 0$ is a real parameter describing essentially the quotient of diffusivities for $x_{i,1}$ and $x_{i,2}$ .

By definition of $\lambda$, all stationary solutions of system (8.1) are independent of $\lambda$. Thus any stationary solution $x$ gives rise to a "trivial branch" $\lambda \mapsto (\lambda, x)$ of stationary solutions. In particular, there is a unique **homogeneous** stationary solution $x^0 = (x_1^0, \dots, x_n^0)$, i.e. a stationary solution with $x_1^0 = \dots = x_n^0$. This solution is given by

$$x_1^0 = \cdots = x_n^0 = (1, b). \tag{8.3}$$

The associated trivial branch is

$$\lambda \mapsto (\lambda, x^0).$$

Before going into $D_n$-equivariance of system (8.1), let us briefly analyse the eigenvalues $\mu$ of the linearization $L$ of (8.1) at this homogeneous stationary solution $(\lambda, x^0)$. This linear analysis is quite similar to Turing's, cf. [Tu]. Note that $\mu \in \mathbb{C}$ is an eigenvalue of $L$ if and only if for some $\mu_{\tilde{\Delta}} \in spec(-\alpha^2\tilde{\Delta})$

$$det\left(\mu\begin{pmatrix}1 & 0\\ 0 & \lambda\end{pmatrix} + \mu_{\tilde{\Delta}}\cdot id - B\right) = 0. \tag{8.4}$$

Here $spec(-\alpha^2\tilde{\Delta})$ denotes the set of real eigenvalues of

$$-\alpha^2\tilde{\Delta}: \quad \mathbb{R}^{2n} \to \mathbb{R}^{2n},$$

and $B$ is the linearization of $\tilde{f}$

$$B := \begin{pmatrix} b-1 & 1 \\ -b & -1 \end{pmatrix} = D_u\tilde{f}(1,b).$$

In fact, the algebraic multiplicity of an eigenvalue $\mu$ of $L$ equals the sum of algebraic multiplicities of $\mu$ as a solution of (8.4), summing over those $\mu_{\tilde{\Delta}} \in spec(-\alpha^2\tilde{\Delta})$ for which (8.4) holds. Indeed $\alpha^2\tilde{\Delta}$ commutes with the linearization of the nonlinear part of (8.1) at $x^0$, which is a block diagonal matrix with $n$ identical $2\times 2$ blocks $B$. This proves the above claims.

Writing out the characteristic equation (8.4) explicitly, we obtain

$$\lambda\mu^2 + \left((\mu_{\tilde{\Delta}}+1-b)\lambda + (\mu_{\tilde{\Delta}}+1)\right)\mu + (\mu_{\tilde{\Delta}}+1/\mu_{\tilde{\Delta}}+2-b)\mu_{\tilde{\Delta}} = 0. \tag{8.5}$$

Thus $(\lambda, x^0)$ is a center if and only if

$$\lambda = \frac{\mu_{\tilde{\Delta}}+1}{b-1-\mu_{\tilde{\Delta}}} > 0 \quad \text{and} \tag{8.6.a}$$

$$(\mu_{\tilde{\Delta}}+1/\mu_{\tilde{\Delta}}+2-b)\mu_{\tilde{\Delta}} > 0. \tag{8.6.b}$$

At each such center, the purely imaginary pair $\mu(\lambda), \bar{\mu}(\lambda)$ crosses the imaginary axis from left to right as $\lambda$ increases because

$$\frac{d}{d\lambda}Re\,\mu(\lambda) = \frac{\mu_{\tilde{\Delta}}+1}{2\lambda^2} > 0. \tag{8.7}$$

With this elementary eigenvalue analysis in mind, let us now explore the symmetry aspect of system (8.1). The dihedral group $D_n$ is generated by a rotation $\vartheta$ and a reflection $\kappa$ such that

$$\begin{aligned} \langle\vartheta\rangle &\cong \mathbb{Z}_n \\ \langle\kappa\rangle &\cong \mathbb{Z}_2 \\ \kappa\vartheta\kappa^{-1} &= \vartheta^{-1}, \end{aligned} \tag{8.8}$$

cf. e.g. [Ser, §5.3]. Equation (8.1) is equivariant with respect to the representation $\rho$ of $\Gamma := D_n$, acting on $x = (x_1, \cdots, x_n) \in \mathbb{R}^{2n}$ by

$$\left.\begin{aligned} (\rho(\vartheta)x)_i &= x_{i-1} \\ (\rho(\kappa)x)_i &= x_{-i} \end{aligned}\right\} , \qquad i (mod\ n). \tag{8.9}$$

To find discrete waves in system (8.1) we apply theorem 2.10. As a cyclic factor $H_0/K_0 \cong \mathbb{Z}_n$ of $\Gamma = D_n$ we take $H_0 := \langle\vartheta\rangle \cong \mathbb{Z}_n$ and $K_0 = \{id\}$. Then we choose some binary orbit $d \in D(n)$, $d \subset \mathbb{Z}/n\mathbb{Z}$, cf. definition 2.4. Let $r \in d \cup (-d)$ denote an element which is closest to 0 (putting $r = 0$ in case $d = \{0\}$). We claim that all assumptions of theorem 2.10 are satisfied if the coefficients $\alpha^2$ and $b$ in system (8.1) are chosen such that

$$1 + 4\alpha^2 \sin^2(\pi r/n) \quad < \quad b \quad < \quad 4. \tag{8.10}$$

Let us first check assumptions (2.20.*a*-*c*). Note that $\Lambda = (0, \infty)$ here, $X = \mathbb{R}^{2n}$, and $X^{H_0} = \{(x_1, \ldots, x_n) \mid x_1 = \cdots = x_n \in \mathbb{R}^2\}$ is the set of homogeneous states $x \in X$. In $X^{H_0}$ there exists a unique stationary branch $(\lambda, x^0)$ given by (8.3). This branch has only nonzero eigenvalues, by the characteristic equation (8.5), because condition (8.10) implies

$$(\mu_{\tilde{\Delta}} + 1/\mu_{\tilde{\Delta}} + 2 - b)\mu_{\tilde{\Delta}} \quad \geq \quad (4 - b)\mu_{\tilde{\Delta}} \quad > \quad 0,$$

in case $\mu_{\tilde{\Delta}} > 0$, whereas in case $\mu_{\tilde{\Delta}} = 0$ we have

$$(\mu_{\tilde{\Delta}} + 1/\mu_{\tilde{\Delta}} + 2 - b)\mu_{\tilde{\Delta}} \quad = \quad 1 \quad > \quad 0.$$

The set of centers in $\Lambda \times X^{H_0}$ is bounded since $\lambda$, defined by (8.6.*a*), is positive for only finitely many $\mu_{\tilde{\Delta}}$. Analytic dependence of the linearization on $\lambda$ is obvious, so that assumptions (2.20.*a*-*c*) indeed hold.

It remains to check the condition

$$\mathcal{H}^{\pm d}_{H_0,K_0} \quad \neq \quad 0 \tag{2.28}$$

on the global equivariant Hopf index $\mathcal{H}$. We recall from definition 2.8 that

$$\mathcal{H}^{\pm d}_{H_0,K_0} \quad = \quad (-1)^{E_0^0(\lambda_0)} \sum_{j \in d \cup (-d)} \chi_0^j \,, \tag{8.11}$$

where the net crossing numbers $\chi_0^j$ are evaluated for $\lambda \in (0, \infty)$ instead of $\lambda \in (-\infty, \infty)$, cf. (2.24) and remark 2.11. By (8.7), eigenvalues $\mu$ can cross the imaginary axis only from left to right as $\lambda$ increases. Hence the $\chi_0^j$ are all nonnegative, and $\chi_0^j$ is positive if any center with representation $j$ occurs at all. We claim

$$\chi_0^r \quad > \quad 0; \tag{8.12}$$

then $\mathcal{X} \neq 0$ is immediate.

To find a center with representation $r$ of $H_0 = \langle\vartheta\rangle \cong \mathbb{Z}_n$ we decompose $X^{K_0} = X = \mathbb{R}^{2n}$ into representation subspaces

$$X \quad = \quad \bigoplus_{0\leq j\leq n/2} X_j$$

as in (2.22). Using e.g. [Ser, §2.6] we find that $X_r$ is spanned by the real and imaginary parts of the vectors

$$\sum_{k=1}^{n} exp(-2\pi ikr/n)\cdot e_{k,\iota} \quad , \qquad \iota = 1,2\,, \tag{8.13}$$

with $e_{k,\iota}$ denoting the unit vectors in $\mathbb{R}^{2n}$. A short calculation shows that the restriction to $X_r$ of

$$-\alpha^2\tilde{\Delta} \quad = \quad -\alpha^2\Big(\rho(\vartheta)^{-1} - 2\cdot id + \rho(\vartheta)\Big)$$

acts as multiplication by the eigenvalue

$$\mu_{\tilde{\Delta}} \quad = \quad 4\alpha^2\sin^2(\pi r/n)\,. \tag{8.14}$$

Going back to condition (8.6.a) for a center, we indeed obtain a center with $\lambda > 0$ in the representation space $X_r$ because

$$b-1-\mu_{\tilde{\Delta}} \quad = \quad b-1-4\alpha^2\sin^2(\pi r/n) \quad > \quad 0$$

by condition (8.10) on $\alpha^2$ and $b$. This proves $\chi_0^r > 0$ (and in fact $\chi_0^r = 2$ if $r \neq 0, n/2$). Consequently

$$\mathcal{X}^{\pm d}_{H_0,K_0} \quad \neq \quad 0\,.$$

Thus theorem 2.10 applies to the ring of $n$ coupled Brusselators and we obtain a global continuum

$$\mathcal{C} \quad \subseteq \quad (0,\infty)\times X$$

of discrete waves. We claim that theorem 2.10 actually controls virtual symmetry on $\mathcal{C}$ to be given precisely (not just "at least") by

$$(H_0, K, \Theta)\,, \quad \Theta \in d \cup (-d)\,,$$

in case $d \neq \{0\}$. Indeed, suppose that $(H_0, K, \Theta)$ extends to a larger symmetry $(\hat{H}, \hat{K}, \hat{\Theta})$. Then necessarily $\hat{H} = D_n = \Gamma$, because $\Gamma$ itself is the only group strictly containing $H_0$. Because $\hat{H}/\hat{K}$ is cyclic, this implies $\hat{K} \geq \mathbb{Z}_n = H_0$. Hence $K = \hat{K} \cap H_0 = H_0$, i.e. $\Theta = 0$, $d = \{0\}$. This contradicts $d \neq \{0\}$ and hence proves our claim. Periodic solutions with $\Theta = 0$ are homogeneous

$$x_1(t) \quad = \quad \ldots \quad = \quad x_n(t)\,.$$

Their symmetry is given by

$$(D_n, D_n, 0)\,.$$

Let us return to a nonhomogeneous periodic solution $x(t)$ with virtual symmetry $(H_0, K, \Theta)$, $\Theta \neq 0$. Assume $x(t)$ has no virtual period except for the minimal period itself. Then $(H_0, K, \Theta)$ is the symmetry of $x(t)$, by lemma 4.12. In particular this holds if $x(t)$ is hyperbolic. Now we have reached our goal, set up in question (1.7): we have found periodic solutions with prescribed symmetry, in a global sense.

In accordance with remark 2.11 we note that the continuum $C$ is also called global if it stays bounded and just extends to arbitrarily small values of $\lambda > 0$. For system (8.1), however, this phenomenon should not occur. We sketch a reason. If $\lambda > 0$ is very small, then $x(t)$ rapidly approaches the set

$$x_{i,1}^2 x_{i,2} \quad = \quad b x_{i,1} \qquad \text{for all } 0 < i \leq n ,$$

to within a small distance (using singular perturbation analysis, see e.g. [Hopp]). Afterwards, the dynamics is equivalent to that of

$$\dot{x}_{i,1} \quad = \quad \alpha^2 (x_{i-1,1} - 2x_{i,1} + x_{i+1,1}) \; + \; 1 \; - \; x_{i,1}$$

which tends to the equilibrium $x \equiv x^0$. Therefore $C$ is unbounded, or contains arbitrarily large virtual periods.

As a very concrete example we pick $n = 1986 = 2 \cdot 3 \cdot 331$. How many global equivariant Hopf indices $\mathcal{H}^{\pm d}_{H_0,K_0}$ do we get for $H_0 = \mathbb{Z}_n$ and $K_0 = \{id\}$? Equivalently: how many different sets $d \cup (-d)$ occur? At the end of §2.3 we have found

$$|D(n)| \quad = \quad 24 \qquad \text{for } n = 1986 \; .$$

Observing that $d = -d$ for any $d \in D(n)$, we thus obtain 24, possibly different, global equivariant Hopf indices for 1986.

Let us compare this example, at least phenomenologically, to the result of Alexander & Auchmuty [Ale&Au2] mentioned in §1.2. For a more in-depth discussion see §9.4. Alexander & Auchmuty assume some "oddness" for $\chi_r$, roughly, and obtain global continua (in a somewhat different sense) of periodic solutions $(\lambda, x(t), \tilde{p})$ with

$$x_{i+1}(t) \quad = \quad x_i(t + (l/n) \cdot \tilde{p}) \, , \tag{8.15}$$

where $\tilde{p} = kp$ is some multiple of the unknown minimal period $p$ of $x(t)$. Identifying $H_0$ with $\mathbb{Z}_n$ via $\vartheta = 1/n$ such solutions have symmetry $(H_0, K, \Theta)$, in our notation, with

$$\Theta \quad \equiv \quad -kl \qquad\qquad (mod\ n) \tag{8.16}$$

because

$$\begin{aligned} x_i(t + (kl/n)p) \quad &= \quad x_{i+1}(t) \quad = \quad \Big(\rho(1/n)\, x(t - (\Theta/n)p)\Big)_{i+1} \quad = \\ &= \quad x_i(t - (\Theta/n)p) \, . \end{aligned}$$

For example, pick some $l$ which is relatively prime to $n$. Because Alexander & Auchmuty have no control on the minimal periods $p$ of the solutions, $k$ may be anything. In particular, $\Theta$ can be anything, by (8.16). In contrast, our results retain some control over $\Theta$ as we have illustrated with the 24 different Hopf indices for $n = 1986$. On the other hand, if $n$ is a prime number with maximal multiplicative order of 2,

$$ord_n^*(2) \quad = \quad n - 1 ,$$

then $|D(n)| = 2$ and we do not gain much more control over $\Theta$ than Alexander & Auchmuty have already. This is expected to happen for about 37% of all prime numbers by Artin's conjecture [Sha, §32]. In these "worst" cases, our condition $\mathcal{N} \neq 0$ is only slightly better than a condition like $\mathcal{N}$ odd.

Picking $H_0 = \mathbb{Z}_n \leq D_n$, so far, we have not encountered reflection invariant waves like $H = K = \langle\kappa\rangle$, or $\langle\vartheta\kappa\rangle$, which are sometimes called standing waves. In that case, our global results are much weaker, and are more reminiscent of symmetries in stationary bifurcation. In fact, such waves can bifurcate into $H = \langle\kappa\rangle, K = \{id\}$ through a flop, cf. §5. And if $H = K = \langle\kappa\rangle$, then the actual virtual symmetry may be much bigger than just $(\langle\kappa\rangle, \langle\kappa\rangle, 0)$. For example it might be given by

$$(D_{\tilde{n}}, D_{\tilde{n}}, 0) ,$$

where $\tilde{n}$ is a divisor of $n$. Spreading some optimism: a generic theory for global $D_n$-equivariant Hopf bifurcation seems feasible and might help here.

We conclude our discussion of Turing rings with a quote from Turing's original paper [Tu, §13], written in 1951. The term *"wave" theory* basically stands for his analysis of the linearized equations. Turing writes:

*The "wave" theory which has been developed here depends essentially on the assumption that the reaction rates are linear functions of the concentrations, an assumption which is justifiable in the case of a system just beginning to leave a homogeneous condition. Such systems certainly have a special interest as giving the first appearance of a pattern, but they are the exception rather than the rule. Most of an organism, most of the time, is developing from one pattern into another, rather than from homogeneity into a pattern. One would like to be able to follow this more general process mathematically also. The difficulties are, however, such that one cannot hope to have any very embracing* **theory** *of such processes, beyond the statement of the equations. It might be possible, however, to treat a few particular cases in detail with the aid of a digital computer.*

## 8.3 Reaction diffusion systems

A Brusselator reaction diffusion system with $O(3)$ symmetry is given by

$$\begin{aligned} \begin{pmatrix} 1 & 0 \\ 0 & \lambda \end{pmatrix} x_t &= \alpha^2 \Delta x + \tilde{f}(x), & (\xi \in \Omega) \\ \partial_\nu x &= 0 & (\xi \in \partial\Omega)\,. \end{aligned} \tag{8.17}$$

Here the domain $\Omega$ denotes the unit ball in $\mathbb{R}^3$, $x = x(t, \xi) = (x_1, x_2)(t, \xi) \in \mathbb{R}^+ \times \mathbb{R}^+$ are concentrations, $\tilde{f}(x)$ is again the Brusselator kinetics (8.2), $\lambda > 0$ is a real parameter describing essentially the quotient of diffusivities for $x_1$ and $x_2$, and $\alpha^2 > 0$ is a diffusion constant. By standard theory [Hen], assumptions (2.32.$a, b$) are satisfied on the Sobolev space

$$X = H^2(\Omega) \times H^2(\Omega) .$$

Therefore (8.17) defines an analytic semigroup on $X$ and, by corollary 2.13, we can apply theorems 2.9 and 2.10 to this reaction diffusion system, once conditions (2.32.$c, d$) on the group action are checked.

As before, we note that the stationary solutions are independent of the parameter $\lambda > 0$. There is a unique homogeneous, i.e. $\xi$-independent, stationary solution

$$x^0 \equiv (1, b) \tag{8.18}$$

with associated trivial branch $(\lambda, x^0)$. This is due to Neumann boundary conditions. The linearized stability analysis along the trivial branch now proceeds verbatim as in the coupled oscillator case, i.e. equations (8.4) – (8.7) remain valid, if we just replace $\mu_{\tilde{\Delta}}$ by $\mu_\Delta \in spec(-\alpha^2\Delta)$ everywhere. Indeed we obtain (8.17) from (8.1) by just replacing the discretized Laplacian $\tilde{\Delta}$ by the Laplacian $\Delta$ with Neumann boundary conditions.

Let us now consider the symmetry aspect of the reaction diffusion system (8.17). The system is equivariant with respect to the representation $\rho$ of $\Gamma = O(3)$ on $x \in X$ given by

$$(\rho(\gamma)x)(\xi) \quad := \quad x(\gamma^{-1}\xi)\,. \tag{8.19}$$

Indeed $\Delta$ with Neumann boundary conditions on the unit ball $\Omega$ commutes with this representation, and so does the nonlinearity $\tilde{f}$. Therefore equivariance assumption (2.32.*c*) holds. For a general background on $\Delta$, spherical harmonics, representations of $O(3)$ and local bifurcations we refer to [Go&Sch&St, Go&St1, Ihr&Go, Smo&Was3, Van5]. Taking the domain $\Omega$ to be a ball provides a special case of an axisymmetric domain. Global Hopf bifurcation for axisymmetric domains (rotating and frozen waves) was investigated by Alexander &Auchmuty [Ale&Au1], and we comment on their results at the end of §8.3.1. For a numerical simulation in a disk see [Ern&H-K] and, for an experiment, fig.1.2. For another numerical example on the circle see [Doe&Ker2].

**8.3.1.** Let us try to find **rotating waves** in system (8.17). We need a cyclic factor $H_0/K_0 \cong \mathbb{Z}_\infty = \mathbb{R}/\mathbb{Z}$ of $\Gamma = O(3)$. Such a factor is given by $H_0 = SO(2)$ and $K_0 = \{id\}$. We postpone a discussion of other choices of $H_0, K_0$, for the moment. Note that $\mathcal{R}$ is given by an expression

$$\mathcal{R}x \quad = \quad 2\pi\,(-\xi_2\,\partial_{\xi_1}x \;+\; \xi_1\partial_{\xi_2}x)\,,$$

in suitable coordinates $\xi = (\xi_1, \xi_2, \xi_3)$. Thus $\mathcal{R} : X_\omega \to X$ is bounded if we pick $\omega > 1/2$. In other words, assumption (2.32.*d*) holds and theorem 2.9 can be applied.

We select any nonzero binary orbit $d \in D(\infty)$, cf. table 2.2:

$$d \quad = \quad \{2^j r \mid 0 \le j < \infty\}, \qquad r \text{ odd}\,. \tag{8.20}$$

Let $\mu_\Delta^r$ denote the smallest eigenvalue of the radial boundary value problem

$$\begin{aligned} u''(R) + \frac{2}{R}u'(R) - \frac{r(r+1)}{R^2}u(R) \quad &= \quad -\,\mu_\Delta^r\,u(R) \\ u'(0) \quad = \quad u'(1) \quad &= \quad 0\,. \end{aligned} \tag{8.21}$$

Note that $\sqrt{\mu_\Delta^r}$ is the first positive maximum of the spherical Bessel function $j_r$, see [Abram& Ste, ch.10]. We claim that all assumptions of theorem 2.9 are satisfied if the coefficients $\alpha^2$ and $b$ in system (8.17) are such that

$$1 + \alpha^2\mu_\Delta^r \quad < \quad b \quad < \quad 4\,. \tag{8.22}$$

Checking assumption (2.20.*a*) we immediately notice a pitfall. Even $X^\Gamma = X^{O(3)}$ may contain many stationary solutions, possibly degenerate, besides the trivial branch $(\lambda, x^0)$ of homogeneous stationary solutions. Indeed any radial solution lies in $X^\Gamma$, and any axisymmetric solution lies in $X^{SO(2)}$, up to conjugacy. So we will follow remark 2.11 and apply theorem 2.9 only in the subset $\mathcal{Y}$ of $\Lambda \times X$ which has all axisymmetric stationary solution branches removed, except for the trivial branch $(\lambda, x^0)$ itself. Then assumption (2.20.$a$) holds, as in the $D_n$ case. Similarly assumption (2.20.$b$) persists to hold for the reaction diffusion system (8.17).

Again it only remains to check the condition

$$\mathcal{H}^{\pm d}_{H_0,K_0} \quad \neq \quad 0 \tag{2.28}$$

on the global equivariant Hopf index (definition 2.8). By the same reasoning as in the $D_n$ case it is sufficient to show that some center with the nontrivial representation $r$ of $SO(2)$ occurs. In the light of our assumption (8.22) and of condition (8.6.$a$) for a center it is therefore sufficient to show that

$$\mu_\Delta := \alpha^2 \mu^r_\Delta \quad \in \quad spec(-\alpha^2 \Delta), \qquad \text{and} \tag{8.23.a}$$

the representation $r$ of $SO(2)$ occurs in the eigenspace of $\mu_\Delta$ . (8.23.$b$)

Writing out $\Delta$ in spherical coordinates, (8.23.$a$) is immediate. Moreover, the representation of $O(3)$ on the eigenspace of $\mu_\Delta$ comes from the representation of $O(3)$ on the space $V_r$ of spherical harmonics of degree $r$; $dim\ V_r = 2r+1$. Of course, $SO(2)$ acts on $V_r$. The decomposition of $V_r$ into irreducible representations of $SO(2)$ is given by

$$V_r \quad = \quad W_0 \oplus W_1 \oplus \ldots \oplus W_r \tag{8.24}$$

where $dim\ W_0 = 1$, $dim\ W_k = 2$ for $k > 0$. This is the Cartan decomposition of the irreducible representation $V_r$ of $O(3)$ into irreducible representations of the maximal torus $SO(2)$. Knowing spherical harmonics, this decomposition can be obtained explicitly; see e.g. [Brö&tD, §II.10, Go&Sch&St, §XIII]. In the decomposition (8.24), the group $SO(2)$ acts by representation $k$ on $W_k$. Putting $k = r$ this proves (8.23.$b$), and hence $\mathcal{H}$ is nonzero.

Thus theorem 2.9 applies to the Brusselator system (8.17) in the unit ball and we obtain a global continuum

$$\mathcal{C} \quad \subseteq \quad \mathcal{Y} \quad \subseteq \quad (0,\infty) \times X$$

of rotating and frozen waves. Note that $\mathcal{C}$ might terminate at a nonhomogeneous axisymmetric stationary solution because such solutions are not contained in $\mathcal{Y}$. By theorem 2.9 the only other possibility is that $\mathcal{C}$ is unbounded in $\Lambda \times X$. In fact we do not expect $\mathcal{C}$ to extend down to $\lambda = 0$, by the same singular perturbation reasoning as in §8.2. Note that our solutions have virtual symmetry at least

$$(SO(2),\, K, \Theta)\,, \quad \Theta = \pm 2^j r \quad \text{for some } j\,. \tag{8.25}$$

Solutions with such symmetry may be visualized as rotating or frozen spiral waves with $2^j r$ arms. Strictly speaking, however, our symmetry analysis does not distinguish spirals from other geometric objects. Also, we only count identical arms as contributing to $2^j r$: the symmetry point of view may amount to a "numerical arms reduction".

Agladze & Krinsky [Agl&Kri, Kri&Agl] have created spiral waves with 1 to 4 arms in a Belousov-Zhabotinskii reaction, experimentally. They use a setting with $O(2)$ geometry, which can be treated analogously to the above lines. In [Kri&Agl, §4.1], a relation between $\Theta$ (called "topological charge $N$" there) and minimal period $p$ was described as follows:

*The period of a spiral wave $S_N$ was experimentally found to increase with the topological charge (for $|N| \leq 4$). Thus it might be expected that spiral waves with a lower topological charge can synchronize those with a greater topological charge.*

The flop doubling of rotating waves is a theoretical example for such a synchronization, since $\Theta$ jumps by a factor two. However, $\alpha = (\Theta p)^{-1}$ varies continuously. Therefore, in contrast to the above observation, minimal period decreases with the "topological charge" $\Theta$ at a flop doubling. Experimentally, on the other hand, these spirals are not close to each other in phase space $X$. Moreover, they do not seem to appear in a rigidly rotating fashion – we comment on a possible explanation in §9. Numerical simulations showing several arms were performed by Rovinsky [Rov], unfortunately in a $D_4$ geometry ($\Omega$ is a square) which somewhat resists a rotating wave analysis. For further discussion see also the survey [Win] by Winfree.

Can the symmetry

$$(SO(2), K, \Theta)\,, \quad \Theta = \pm 2^j r \tag{8.25}$$

extend to a larger virtual symmetry? Let $\mathbb{Z}_2^c := \{id, -id\} \leq O(3)$. Then the symmetry above could extend to

$$(\mathbb{Z}_2^c \oplus SO(2),\, \mathbb{Z}_2^c \oplus K,\, \Theta) \tag{8.26}$$

in the obvious way. This extension corresponds to solutions $x$ with

$$x(t, \xi) \quad = \quad x(t, -\xi)\,.$$

Checking a list of subgroups of $O(3)$, as given in [Ihr&Go,theorem 2.8, Go&Sch&St,§XIII], for example, we see that (8.25), (8.26) with $K$ finite exhaust all possibilities for $\mathbb{Z}_\infty$ factors in $O(3)$, up to conjugacy. Alternatively, considering the Weyl group of the maximal torus $SO(2)$ in $O(3)$ leads to the same conclusion. Thus (8.26) cannot be extended if $\Theta \neq 0$, and (8.26) is the only possible extension of (8.25). Picking $H_0 = \mathbb{Z}_2^c \oplus SO(2), K_0 = \mathbb{Z}_2^c$ we can proceed analogously to the above discussion to obtain rotating waves with virtual symmetry (8.26). We only have to replace condition (8.22) by the slightly stronger version

$$1 + \alpha^2 \mu_\Delta^{(r+1)} \quad < \quad b \quad < \quad 4\;,$$

because $-id \in O(3)$ acts as $(-1)^r$ on $V_r$ when $r$ is odd, while $-id \in O(3)$ acts trivially on $V_{r+1}$. Let us also recall again that the word "virtual" can be dropped e.g. for hyperbolic rotating waves, by lemma 4.12.

Comparing our results to those of Alexander & Auchmuty [Ale&Au1], our remarks of §8.2 still apply. Analogously to §5.3, they solve

$$-\alpha \mathcal{R} x \,+\, A(\lambda)x \,+\, f(\lambda, x) \quad = \quad 0 \tag{5.3}$$

and obtain a global continuum of triples $(\alpha, \lambda, x)$ under some "oddness" condition on the crossing numbers $\chi_r$. This yields rotating, frozen, or concentric waves – but no control on $\Theta$. At the expense of introducing virtual symmetry we can recover and extend their result.

**8.3.2.** Next we turn to **tetrahedral waves**. We choose $H_0 = \mathbb{Z}_2^c \oplus T$ and $K_0 = \mathbb{Z}_2^c \oplus D_2$. Here $T \leq SO(3)$, called the tetrahedral group, describes the symmetry of a regular tetrahedron ( $T \cong A_4$ is the group of even permutations of 4 elements given by the vertices of the tetrahedron). As a subgroup, $T$ contains $D_2$ which consists of rotations over $\pi$ around the three axes joining opposite mid-edges of the tetrahedron ($D_2 \cong \mathbb{Z}_2 \oplus \mathbb{Z}_2$ is the Klein 4-group). We note that

$$H_0/K_0 \quad \cong \quad \mathbb{Z}_3$$

is cyclic. For a reference see [Ihr&Go, Go&St1]. In [Go&St1] local Hopf bifurcation with tetrahedral symmetry was observed for the first time.

We pick the nonzero binary orbit

$$d \quad = \quad \{\pm 1\} \quad \in \quad D(3)$$

and assume that condition (8.22) holds for $r = 2$, i.e.

$$1 + \alpha^2 \mu_{\Delta}^{(2)} \quad < \quad b \quad < \quad 4 . \tag{8.22$''$}$$

As usual, conditions (2.20.$a$-$c$) then hold with $\Lambda \times X$ replaced by $\mathcal{Y}$ (i.e. all nonhomogeneous stationary branches in $X^{H_0}$ are removed). Moreover

$$\mathcal{X}^{\pm d}_{H_0, K_0} \quad \neq \quad 0 ,$$

because a center occurs with a representation $r \in d$ of $H_0/K_0 \cong \mathbb{Z}_3$. We claim that such a center comes from the representation of $O(3)$ on the five-dimensional space $V_2$ of spherical harmonics of order 2. Indeed we obtain from [Go&St1,§15, Go&Sch&St,§XIII] the following dimensions of fixed point subspaces

$$\begin{aligned} dim \quad V_2^T \quad &= \quad 0 \\ dim \quad V_2^{D_2} \quad &= \quad 2 . \end{aligned}$$

Thus $T/D_2 \cong \mathbb{Z}_3$ acts nontrivially while $\mathbb{Z}_2^c$ acts trivially on $V_2$, and our claim can be proved along the previous lines.

Checking the list of subgroups of $O(3)$ [Ihr&Go, theorem 2.8] we see that such symmetries

$$(\mathbb{Z}_2^c \oplus T, \, \mathbb{Z}_2^c \oplus D_2, \, \Theta) , \quad \Theta \in \{\pm 1\} \tag{8.27}$$

cannot be extended. Thus we obtain a global continuum $\mathcal{C} \subseteq \mathcal{Y}$ of tetrahedral waves with virtual symmetry given by (8.27). Again we do not expect $\mathcal{C}$ to extend down to $\lambda = 0$.

In the special setting of our reaction diffusion system (8.17) it was natural to give the weakest condition, (8.22)$''$, which still guarantees bifurcation of tetrahedral waves. This condition involves the representation of $O(3)$ on $V_2$, the spherical harmonics of degree 2. However, tetrahedral waves also bifurcate from all centers which involve representations on $V_r$, $r \geq 4$. Their symmetry (no "at least", no "virtual" this time ) is given, locally, by

$$(\mathbb{Z}_2^c \oplus T, \, \mathbb{Z}_2^c \oplus D_2, \, \Theta), \quad \Theta \in \{\pm 1\} \qquad \text{for even} \quad r \geq 4 , \tag{8.28}$$

$$(\mathbb{Z}_2^c \oplus T, \, D_2, \, \Theta), \quad \Theta \in \{\pm 1, \pm 2\} \qquad \text{for odd} \quad r \geq 4 . \tag{8.29}$$

Using the language of Golubitsky & Stewart [Go&St1], the reason is the following. For the action of $\Gamma \times S^1$ on the eigenspaces of purely imaginary eigenvalues associated to $V_r$, the symmetries (8.28) resp. (8.29) turn out to be maximal isotropy subgroups for each $r \geq 2$, $r \neq 3$. In our discussion in §9.2 we will show that there is always bifurcation, locally, for any maximal isotropy subgroup (theorem 9.1). This proves our claim in the example. In our case, the dimensions of fixed point subspaces belonging to (8.28), (8.29) are at least two. In contrast, [Go&St1, §15] compute those $r$ for which the dimensions of these isotropy subspaces are exactly two. Therefore they get bifurcation of tetrahedral waves only from $V_r$ with $r = 2, 4, 5, 6, 7, 9$. In this respect, our global results are stronger even when applied only locally.

## 8.4 More examples

In this section we indicate various examples to which our theory applies. However, we do not elaborate on any of these examples. We only describe the natural symmetry $\Gamma$ of the problem and, occasionally, give candidates for $H_0, K_0, H_0/K_0$. These choices suggest certain types of rotating or discrete waves which may relate to observed phenomena. To corroborate the suggestions it is necessary to first select suitable one-parameter models of these phenomena. Then one can compute the global Hopf index as in the previous examples. This can be done, in principle, and it may require numerical analysis in some of the problems described below. Surveying our examples, we also collect hints on open questions, which we discuss in the next section.

**8.4.1.** Our first example is theoretical and not too specific. Centered around the paper by Golubitsky and Stewart [Go&St1] there are many local results which are related to applications; see e.g. [Go&St2, Go&St3, Mon&Rob&St, Rob&Swi&Wag]. A fair amount of work is spent on finding, for given representations, isotropy subgroups of $\Gamma \times S^1$ with two-dimensional fixed point subspace. As in the example of tetrahedral waves above (§8.3.2) this dimension restriction is not necessary to prove local Hopf bifurcation for maximal isotropy subgroups. This will be established in theorem 9.1 below.

**8.4.2.** A second, still general class of examples are Hamiltonian systems

$$\begin{aligned} \dot{p} &= -H_q(p,q) \\ \dot{q} &= H_p(p,q) \end{aligned} \tag{8.30}$$

with $\Gamma$-invariant Hamiltonian $H(\gamma p, \gamma q) = H(p,q)$, for all $\gamma \in \Gamma$. For a background and more general settings see e.g. [Ab&Mars]. Special examples are the spherical pendulum, $\Gamma = O(2)$, and a circular ring of Hamiltonian coupled oscillators, $\Gamma = D_n$ [vG&Val]; see also [vdM]. Adding dissipation, (8.30) becomes a one-parameter family

$$\begin{aligned} \dot{p} &= -H_q(p,q) + \lambda H_p(p,q) \\ \dot{q} &= H_p(p,q) + \lambda H_q(p,q) \end{aligned} \tag{8.31}$$

and theorems 2.9, 2.10 yield a global version of the Ljapunov center theorem with symmetry. For a detailed local analysis see [vdM, Mon&Rob&St]. We note that $\mathcal{N} = 0$ in the $1 : -1$ resonance analysed in [vdM]. In the case of no symmetry, the dissipation trick was already used in [Ale&Y1].

**8.4.3.** A famous example, which contains symmetry for conceptual simplification, is Eigen's hypercycle [Ei&Schu, Hof&Sig]. It consists of a circular loop of sequentially reacting enzymes. The symmetry is $\Gamma = \mathbb{Z}_n$ and we may take $H_0 = \mathbb{Z}_n$ and $K_0 = \{0\}$. Waves with such symmetries might be interpreted as generators for biological clocks; symmetry breaking (flip, flop) is related to frequency doubling. A local Hopf bifurcation analysis, disregarding the symmetry aspect, was performed in [Hub].

**8.4.4.** The example of a Turing ring, §8.2, can be generalized as follows. Consider a finite, directed or undirected graph $\mathcal{G}$, i.e. a set of vertices together with a neighboring relation which defines the edges. Given $i \in \mathcal{G}$ let $\mathcal{N}_i \subseteq \mathcal{G}$ denote the set of those neighbors of $i$ which connect to $i$. Let each vertex be occupied by an "oscillator", i.e. by some ODE kinetics

$\tilde{f}(x_i)$, $\tilde{f} \in C^1(\mathbb{R}^N, \mathbb{R}^N)$. Fix a coupling function $c = c(x_j, x_i)$, $c \in C^1(\mathbb{R}^N \times \mathbb{R}^N, \mathbb{R}^N)$ and consider the system

$$\dot{x}_i \quad = \quad \tilde{f}(x_i) + \sum_{j \in \mathcal{N}_i} c(x_j, x_i) - \sum_{j:\ i \in \mathcal{N}_j} c(x_i, x_j), \qquad i \in \mathcal{G}. \tag{8.32}$$

For similar settings see e.g. [Ash&Oth, Stu&Marek]. Disregarding symmetry aspects, global Hopf bifurcation for such networks has been studied by Alexander [Ale2].

Let $\Gamma$ be the symmetry group of the graph, i.e. the set of permutations $\gamma$ of $\mathcal{G}$ such that $\mathcal{N}_{\gamma(i)} = \gamma(\mathcal{N}_i)$, for all $i$. Then (8.32) is equivariant with respect to $\Gamma$. Choosing an appropriate parameter $\lambda$ our global theory applies. Oscillations in crystal lattices come to mind as an example for infinite graphs $\mathcal{G}$. Imposing spatial periodicity this problem reduces to (8.32). Of course, other generalizations are possible which differ from the particular form of (8.32). For global decoupling effects which can lead to toroidal motions, for example, see [Ale&Fie].

Linear coupling $c(x_j, x_i) = x_j - x_i$ in an undirected $n$-gon leads to the Turing ring (8.1) of §8.2. If we consider electric coupling between cells [Cro&Field] then logarithmic expressions for $c$ are more appropriate. Working with convolution type integral equations, instead of ODEs, we may also treat neural nets as in [adH, Fie4].

As a side remark we note here that graphs (8.32) may oscillate stably, even when the individual $\tilde{f}$ does not, see [Sma3, How, Ale3]. This effect occurs already for just two symmetrically coupled oscillators. Two coupled oscillators can also show apparently chaotic dynamics [Ale4, Cro&Field, Schr&Hol&Kub&Marek, Schr&Marek1,2], a phenomenon which we attempt to relate to global Hopf bifurcation briefly in §9.5.

**8.4.5.** Our next example concerns the simplest exothermic reaction $A \to B$ in a porous catalyst pellet which follows the Arrhenius kinetics. See [Aris] for an account of the work of Lee and Luss on oscillations in this system, providing a numerical simulation in one space dimension. For an analysis of the problem without symmetry, including the related problem of isothermal oscillations with Langmuir-Hinshelwood kinetics, see [Fie1, Fie3]. Including symmetry $O(2)$ or $O(3)$ for a disk-shaped wafer or a spherical catalyst pellet, respectively, it should be possible to find effects like rotating hot spots ($H_0 = SO(2)$, $K_0 = \{id\}$). This example will certainly require some numerical input to compute the global equivariant Hopf index $\mathcal{H}$. For a chemical engineering survey and a very nice experimental set-up which demonstrates spatially heterogeneous oscillations see [Raz&Schm, Paw&Schm]. A relation to two-parameter problems is indicated in §9.5.

**8.4.6.** Continuing with PDE examples we mention problems on $S^1$, i.e. with circular symmetry. The simplest such example, with $SO(2)$ symmetry, is the scalar parabolic equation

$$x_t \quad = \quad x_{\xi\xi} + f(x, x_\xi), \qquad \xi \in S^1 = \mathbb{R}/\mathbb{Z}. \tag{8.33}$$

Due to the scalar nature of this equation the dynamics of rotating and frozen waves can be analysed in great detail, including instability of rotating waves and heteroclinic orbit connections between them, see [Ang&Fie]. The methods used there (maximum principles, invariant manifolds, and some topology) are entirely different from our present approach. Still, (8.33) is an illustrative example to visualize our present results on global bifurcation of rotating waves $x = x(\xi - ct)$, because such rotating waves are just periodic solutions of the second order equation

$$cx' \quad = \quad x'' + f(x, x').$$

Another system with $SO(2)$-symmetry is given by the ring laser [Ren&Hak]. Besides rotating waves one also observes periodically "modulated" rotating waves. These correspond to invariant tori bifurcating from the rotating wave, see e.g. [Ren]. For a local analysis of $SO(2)$-symmetric systems, including such tori, see [vG&M-P]. We encounter this bifurcation again in fluid dynamics, cf. 8.4.7, and we revisit it in §9.6.

**8.4.7.** Concerning fluid dynamics we mention Bénard convection and the Taylor-Couette experiment, cf. [Go&Sch&St]. In Bénard convection, one of the observed stationary solutions consists of hexagonal rolls covering the plane [Go&Swi&Kno, Sat1]. In [Rob&Swi&Wag] periodic solutions are discussed in such a hexagonal configuration. The associated equivariance group $\Gamma$ is a semidirect product of the torus $SO(2) \times SO(2)$ with the dihedral group $D_6$. In fact, the torus comes from translations in the underlying hexagonal lattice and $D_6$ is the symmetry of the hexagon. In [Rob&Swi&Wag] the simplest Hopf bifurcation with this equivariance is discussed, locally. It occurs in an eigenspace of real dimension 12. All fixed point subspaces for maximal isotropy subgroups (cf. §9.2) happen to be two-dimensional in that case. In view of remark 8.4.1 above, our global results therefore do not improve on this local bifurcation question. On the contrary, the uniqueness and stability results on the bifurcating branches given in [Rob&Swi&Wag] transcend our method. Only some very weak result on exchange of stability enters into proposition 6.3 on the coupling between center index ⏀ and orbit index $\Phi$. The weakness is that stability implies $\Phi = +1$, but not vice versa. Of course, our global techniques are not confined to those "simple", low-dimensional singularities which can be understood thoroughly after sufficient input of ingenuity. Still, the singularity approach is preferable whenever it is viable.

In the Taylor-Couette experiment on fluid flow between rotating cylinders we encounter $\Gamma = O(2) \times SO(2)$ equivariance, if we look for patterns which repeat periodically with a certain fixed characteristic height. Rotating waves occur with various symmetries $(H_0, K, \Theta)$, e.g. as wavy vortices, twisted vortices, and spiral cells; see [Go&St2] and the references there. In [Go&St2] a simultaneous stationary and Hopf bifurcation is analysed, in real 6-dimensional space. Our previous remarks apply.

**8.4.8.** Going beyond rigidly rotating waves there is experimental evidence for torus bifurcations from rotating waves. See [Go&St2], [Rand] for a mathematical account. In Taylor-Couette flow such tori may figure e.g. as "modulated wavy vortices". One idea to find torus bifurcations of this kind by purely local techniques is discussed in [vG&M-P, §4]. They consider generic $O(2)$-equivariant Hopf bifurcation with the usual 4-dimensional eigenspace. As [Go&St1], they obtain bifurcation of two branches of rotating waves (related by reflection) and of a 2-torus of standing waves (related by rotation). Perturbing the reflection symmetry slightly, two generic $SO(2)$-equivariant Hopf bifurcations appear consecutively. They generate the two rotating wave branches. The 2-tori may persist, bifurcating through a secondary bifurcation from one of the rotating wave branches. Note that this example also illustrates nicely how a generic $SO(2)$-equivariant bifurcation picture arises out of an $O(2)$ situation, in harmony with our presentation in §5. Another idea to study torus bifurcations by local techniques is that of mode interaction. Considering $O(2)$-equivariant Hopf bifurcation, Chossat & Golubitsky & Keyfitz [Chos&Go&Key] have studied the local bifurcation behavior at a center with eigenvalues (modes) $\pm i, \pm i\omega$ and $\omega$ irrational. In normal form, they obtain rotating waves, bifurcating 2-tori and even 3-tori, already in the case of a 6-dimensional eigenspace. For more results on mode interactions leading to tori see e.g. [Chos&Dem&Io, Da1, Da2, Da&Arm, Da&Kno, Go&St2], and the survey in [Go&Sch&St]. Of course, one would like to analyse such tori not only near steady states. Let us also revisit here the experiments of Agladze and Krinsky [Agl&Kri, Kri&Agl] on multiarmed spirals in the Belousov-Zhabotinskii medium. The repeated connection and disconnection of arms in the core of the spiral, which they report, might be due to a torus bifurcation from a rigidly rotating multiarmed spiral which has already lost stability. We will return to the torus problem from a theoretical point of view in §9.6.

Concerning tori, let us finish this section with a surprising example due to J. Swift [Swi]. The tori which we have discussed so far all come from rotating waves, i.e. from an $SO(2)$-factor in the equivariance group $\Gamma$. In contrast, Swift considers a Turing ring of 4 coupled oscillators, as in §8.2 above, with $D_4$ symmetry. He detects some open set of one-parameter bifurcation problems in this context, for which a 2-torus bifurcates locally, directly from the stationary solution. This torus occurs due to only a discrete symmetry $D_4$! In a much more degenerate situation, such a possibility has also been observed by Alexander & Auchmuty [Ale&Au2]. Finding such tori certainly lies far beyond the horizon of our present theory.

# 9 Discussion

## 9.1 Outline

We finally put our results into a variety of perspectives. In §9.2 we draw a local conclusion from our global results: locally, there is always bifurcation for any "maximal isotropy" group $H^{\Theta}$ (theorem 9.1). Then we discuss, or rather circumvent, stability questions (§9.3). We believe that this question, if asked globally, should be addressed numerically but we see some obstacles. Anyhow, generic results may prove to be a useful tool. Returning to results of Alexander, Auchmuty, and Yorke on global Hopf bifurcation in §9.4, we compare their topological approach with the genericity, singularity theory approach followed here. Basically we find the genericity approach more flexible, at least for periodic solutions. For example, it is closer to the goal of finding continua with arbitrarily large minimal period. Such continua may lead to homoclinic solutions or to openly chaotic phenomena. In §9.5 we link this circle of ideas to the open question (9.7) of homotopy (non-)invariance of our global equivariant Hopf index $\mathcal{H}$. In systems without symmetry, obstruction to homotopy invariance comes from stationary solutions with a multiple eigenvalue zero. Recall that we have excluded eigenvalues zero altogether by nondegeneracy assumption (2.20.$a$). In §9.6 we discuss eigenvalues zero, imposed intrinsically by equivariance. We are led to consider invariant manifolds of solutions, given by group orbits. This is a generalization of rotating/frozen waves. For an example, we return to the question of 2-tori, which bifurcate from rotating waves. Global bifurcation of such tori is another open question, cf. (9.13), but not beyond the fringes of hope.

## 9.2 Maximal isotropy subgroups

Paradoxically, one might think, global results in fact answer local questions. Here is an example. We consider the $\Gamma$-equivariant $C^1$-system

$$\dot{x} = f(\lambda, x), \tag{9.1}$$

$x \in X = \mathbb{R}^N, \lambda \in \mathbb{R}$ as usual, and assume

$$f(\lambda, 0) = 0; \qquad (9.2.a)$$

$$spec\ D_x f(0,0) = \{\pm i\}, \qquad (9.2.b)$$

where *spec* denotes the spectrum and $i$ is assumed to have geometric (and algebraic) multiplicity $N/2$;

$$sign\ Re(spec\ D_x f(\lambda, 0)) = sign\ \lambda \qquad \text{for } \lambda \text{ near } 0. \qquad (9.2.c)$$

These assumptions are slightly less restrictive than the basic assumptions (H1), (H2) of Golubitsky & Stewart [Go&St1, §5]. Letting $A := D_x f(0,0)$ we have the following action of $\Gamma \times S^1$, $S^1 = \mathbb{R}/\mathbb{Z}$, on $X$

$$\tilde{\rho}(\gamma, \vartheta)\, x_0 := \rho(\gamma)\, e^{-2\pi A \vartheta} x_0\,. \tag{9.3}$$

Note that $(H, K, \Theta)$ is a virtual symmetry of $x = 0$ at $\lambda = 0$ if and only if

$$H^{\Theta} \quad = \quad \{(h, \Theta(h)) \mid h \in H\} \quad \leq \quad \Gamma \times S^1 \tag{9.4}$$

is an isotropy group of some $x \in X \setminus \{0\}$ under this representation, cf. definition 1.2. Moreover $q = 2\pi$ is the only virtual period by assumption (9.2.$b$). For completeness we note that any isotropy group, under $\tilde{\rho}$, of any $x \neq 0$ has the form (9.4) with some homomorphism $\Theta : H \to S^1$.

Following [Go&St1, §12] we call an isotropy group $\Sigma < \Gamma \times S^1$ a **maximal isotropy subgroup** if no other proper isotropy subgroup of $\Gamma \times S^1$ contains $\Sigma$. Assume, for a moment, that an isotropy subgroup $\Sigma$ of $\Gamma \times S^1$ has been found with

$$dim_R X^{\Sigma} \quad = \quad 2\,. \tag{9.5}$$

Then [Go&St1, theorem 5.1] states that local Hopf bifurcation occurs with symmetry $\Sigma = H^{\Theta}$, under slightly stronger assumptions than (9.2.$a$-$c$). Using (9.5), they also obtain uniqueness of the bifurcating local branch. On the other hand, (9.5) implies that $\Sigma$ is a maximal isotropy subgroup, by [Go&St1, proposition 12.5]. Thus [Go&St1] prove local Hopf bifurcation for some maximal isotropy subgroups, namely for those which satisfy dimension restriction (9.5).

But there are maximal isotropy subgroups for which (9.5) does not hold. Tetrahedral waves as discussed in §8.2.2 are an example ($\Gamma = O(3), H = \mathbb{Z}_2^c \oplus T, K = D_2$ or $\mathbb{Z}_2^c \oplus D_2$). The symmetries which we have encountered there are always maximal because $\Theta$ does not extend to groups larger than $H$. And dimension condition (9.5) does not hold for representations which come from spherical harmonics $V_r$ of degrees

$$\begin{array}{lll} r \geq 8, & r & \text{even, or} \\ r \geq 11, & r & \text{odd.} \end{array}$$

I am indebted to Marty Golubitsky, among other things, for asking me the following two questions: "Does local Hopf bifurcation occur for **all** maximal isotropies? If so, why didn't you say it in [Fie5]?" The answer to the first question is "yes". I prefer not to answer the second question.

**9.1 Theorem :**

*Let assumptions (9.2.a-c) hold and let $\Sigma < \Gamma \times S^1$ be a maximal isotropy subgroup for the action (9.3) of $\Gamma \times S^1$.*

*Then $\Sigma = H^{\Theta}$, and there exists a bifurcating (local) continuum $C \subset \Lambda \times X$ with symmetry $(H, K, \Theta)$. More precisely, $C$ contains the center $(\lambda_0, x_0) = (0,0)$, and $C \setminus \{(\lambda_0, x_0)\}$ is nonempty and consists of periodic solutions with minimal period near $2\pi$ and with symmetry $(H, K, \Theta)$.*

**Proof :**

Assumptions (9.2.$a$-$c$) allow us to apply theorem 2.10 on a small neighborhood $\mathcal{Y}$ of $(\lambda_0, x_0) = (0,0)$ in $\Lambda \times X$, cf. remark 2.11. Note that analyticity assumption (2.20.$c$) is not needed here, by assumption (9.2.$c$) and remark 7.2 (§7.3). Assumption (9.2.$c$) further guarantees that, on the small neighborhood $\mathcal{Y}$,

$$\aleph^d_{H,K} \quad \neq \quad 0$$

if we choose the binary orbit $d$ such that $\Theta \in d$. Thus theorem 2.10 yields a global continuum $\mathcal{C} \subseteq \mathcal{Y}$ bifurcating from $(\lambda_0, x_0)$. Any periodic solution $(\lambda, x) \in \mathcal{C}$ has virtual symmetry at least

$$(H, K, \Theta') \quad \text{for some} \quad \Theta' \in d.$$

We claim that $\Theta' = \Theta$ if the neighborhood $\mathcal{Y}$ is chosen small enough. This holds by construction of $\mathcal{C}$ via generic approximations (cf. §7). Indeed, $q = 2\pi$ is the only virtual period of $(\lambda_0, x_0)$ by assumption (9.2.$b$). By virtual period corollary 4.6, this implies the following. Given any small $\varepsilon > 0$ and any large $c > 0$ there exists a neighborhood $\mathcal{Y}$ of $(\lambda_0, x_0)$ and a neighborhood $\mathcal{U}$ of $f$ such that any virtual period $q \leq c$ of any periodic solution contained in $\mathcal{Y}$, for any nonlinearity $g \in \mathcal{U}$, satisfies

$$|q - 2\pi| \quad \leq \quad \varepsilon.$$

Choosing a generic approximation $g_i \to f$, $g_i \in \mathcal{U}$, we recall that a generic continuum $Z_i \subseteq \mathcal{Y}$ bifurcates, for each $i$, as in theorem 2.7. By the remark above, all virtual periods on $Z_i$ deviate from $2\pi$ by at most $\varepsilon$. In particular, $Z_i$ does not contain any period doubling bifurcations, and the minimal periods on $Z_i$ are bounded uniformly by $2\pi + \varepsilon$. Moreover, the symmetry of the periodic solutions on $Z_i$ is always $(H, K, \Theta)$, because symmetry can change only at period doubling bifurcations. Passing to the nongeneric limit, as usual, proves that we may assume $\Theta' = \Theta$ on $\mathcal{C}$.

From $|q - 2\pi| \leq \varepsilon$ we also conclude that virtual period = minimal period for the virtual period $q$ associated to $(H, K, \Theta)$ on $\mathcal{C}$. Thus lemma 4.12 implies, that the (true) symmetry $(\hat{H}, \hat{K}, \hat{\Theta})$ further extends the virtual symmetry, and therefore

$$\hat{H}^{\hat{\Theta}} \quad \geq \quad H^{\Theta} \quad = \quad \Sigma,$$

for any periodic solution $(\lambda, x) \in \mathcal{C}$. Using maximality of $\Sigma$, this implies

$$\hat{H}^{\hat{\Theta}} \quad = \quad H^{\Theta},$$

and the proof is complete. □

For detailed studies of (local) Hopf bifurcation from multiple eigenvalues in the no-symmetry case see [All&Y2, Ize&Mas&Vig, Kie]. These studies suggest results on local Hopf bifurcation with symmetry which would be much more general than theorem 9.1, allowing for resonances etc..

Theorem 9.1 does not address the question of bifurcation of periodic solutions with submaximal isotropy $\Sigma$. In stationary bifurcation problems such submaximal bifurcations can occur, see [Chos, Lau]. We expect similar phenomena in Hopf bifurcation.

## 9.3 Stability and numerics

Admittedly, maximal isotropy subgroups $H^{\Theta}$ with two-dimensional fixed point subspace have the advantage that the bifurcation problem then reduces to a standard Hopf bifurcation which is quite similar to the case of no symmetry. In particular the stability of bifurcating periodic solutions can be calculated, e.g. for $\Gamma = O(2)$ [Go&St1, vG&M-P] and for $\Gamma = D_n$ [Go&St3]; see also [Go&Sch&St]. These computations rely on truncation to polynomial invariants and equivariants, and hence they apply only locally, of course.

A special case, where stability information is fairly complete, is the scalar reaction diffusion equation on the circle

$$x_t = x_{\xi\xi} + f(x, x_\xi), \qquad \xi \in \mathbb{R}/\mathbb{Z}. \tag{9.6}$$

If $x = x(\xi - ct)$ is a rotating wave and if $x_\xi$ has $2j$ zeros, $j \geq 1$, then the strict unstable dimension of $x$ is $2j - 1$ or $2j$, cf. [Ang&Fie]. In particular, all rotating waves are unstable. This also follows from a more general observation for strongly monotone systems, see [Hir2]. Due to the very special (scalar) form of equation (9.6) one can even draw some conclusions on the global dynamic behavior. See [Ang&Fie] for more details.

In general, however, we have to resort to numerical pathfollowing schemes. Such schemes follow solution branches and can, in principle, determine stability of any periodic solution by an eigenvalue decomposition of the Floquet matrix. For systems without equivariance, this has been implemented, see e.g. [Doe&Ker1]. For equivariant systems, however, no general algorithm is available today.

Why? Most pathfollowing schemes (e.g. ALCON, AUTO, BIFPACK, DERPER, PITCON) can deal only with particularly easy bifurcation points like turns, "simple bifurcations", pitchforks, period doublings. See [Deu&Fie&Kun, Doe&Ker1, Kub&Marek, Kü&Mit&Web, Mit&Web, Rhe] and the references there. A practical reason is the abundance of these bifurcations in one parameter applications. A theoretical reason is their genericity in one parameter systems (possibly with $\mathbb{Z}_2$ symmetry). Recently, however, higher singularities (cusps, swallow-tails, "B-points") have attracted more attention [Cu&Sa, Fie3, Fie&Kun1, Fie&Kun2, Jep&Spe1, Jep&Spe2, Kub&Marek, Kun1, Kun2, Roo1, Roo2]. These singularities are tied to multiparameter problems: $\lambda \in \mathbb{R}^j$. Understanding these singularities and their interplay computationally is crucial to efficient pathfollowing in several parameters.

Symmetry is a source of higher singularities, already in one-parameter problems. It causes kernels of linearizations to become ever higher-dimensional. Understanding the relevant local singularities is a prerequisite for a successful numerical analysis. "Relevant" may be replaced by "generic" here. Theorem 5.11 gives a list of such generic singularities for secondary bifurcations of periodic solutions with $\mathbb{Z}_n$ or $\mathbb{Z}_\infty$ symmetry. Surprisingly, maybe, the critical Floquet multipliers are always simple, $\pm 1$. Therefore schemes like AUTO [Doe&Ker1] apply, see e.g. [Doe&Ker2]. This would change already for generic secondary bifurcations with $D_n$ or $O(2)$ symmetry - not to speak of groups with irreducible representations of dimension $> 2$.

Our theoretically-minded approach to higher groups uses generic approximations. They perturb, unfold the contributions from higher symmetry slightly to produce a clean generic diagram which could in principle be tackled by a pathfollowing routine. But this barely hides the difficulty: how much should we perturb, quantitatively? Perturbing too little, the singularity persists numerically. Perturbing too much, the diagram becomes meaningless. This Scylla & Charybdis alternative is reminiscent of unfolding of stationary singularities by PL methods [Peit&Prü]. Viewing the dilemma from a different angle: the full-symmetry singularity should be understood analytically, together with its lower symmetry unfoldings. Indeed numerical discretization tends to introduce such lower symmetry perturbations automatically. It remains to admit that we are far from a comprehensive answer to the problem.

## 9.4 Topology or singularity?

Let us examine the role of singularities and generic approximations in global bifurcation results a little bit more closely. As a preliminary example we revisit global bifurcation of stationary solutions

$$f(\lambda, x) = 0$$

as discussed in §3. We have sketched there how homotopy invariance of Brouwer degree and the global bifurcation result of Rabinowitz can be obtained via generic approximation. Essentially, a generic approximation made 0 a regular value of $f$. If $f$ is equivariant with respect to $\Gamma$, it seems natural to consider $f$ restricted to the spaces $X^K$ for each isotropy subgroup $K$ of $\Gamma$, up to conjugacy, and prove global results on $X^K$ [Cer, Cic, Pos1-3]. This uses the Brouwer degrees $deg^K$ of $f$ restricted to $X^K$, for each isotropy subgroup $K$. Under additional assumptions, the collection of $deg^K$ actually classifies the equivariant homotopy type of $f$ completely. For this equivariant version of the (H.) Hopf theorem see e.g. [tD1, theorem 8.4.1, tD2, ch. II.4] and also the classification in [Hau, theorem 4.5]. For example, the additional assumptions are satisfied for equivariant self-maps of a sphere, if each irreducible representation of $\Gamma$ occurs with multiplicity at least 2. Then the collection of $deg^K$ contains all information on $f$, from a homotopy point of view. Note that, working in $X^K$, we can only get global bifurcation of stationary solutions with isotropy **at least** $K$, similarly to the periodic case. We conclude that the topology and the genericity approach to global stationary bifurcation seem equally powerful.

We consider the case of periodic solutions, but without symmetry, next. We remember that the classical approach of Alexander & Yorke [Ale&Y1] treats global Hopf bifurcation as a two-parameter problem. Equivalently, [Ize1] employs the setting

$$F(p, \lambda, x) = -\dot{x} + pf(\lambda, x) = 0;$$

with parameters $\lambda \in \mathbb{R}$ and $p > 0$ (period, non-minimal) on the space of continuous functions $x(\cdot)$ of period 1. Here 1 need not be the minimal period. They obtain a global bifurcating continuum in $(\lambda, p, x)$ by topological techniques: basically some homotopy theory and the $\mathcal{J}$-homomorphism; see [Ale1] and, for the infinite-dimensional case, also [Ale&Fitz]. The results in equivariant settings by Alexander & Auchmuty [Ale&Au1, Ale&Au2] make use of equivariance explicitly to reduce the equation to a bifurcation problem $F = 0$ with two parameters. Afterwards, the results of Alexander [Ale1] apply and equivariance is ignored. Using equivariance more intrinsically, Matsuoka [Mat, theorem 3] obtains a more general result on local Hopf bifurcation but assuming a free $SO(2)$-action. Phrased in a local setting like §9.2, Matsuoka assumes

$$\sum_{d \in D(\infty)} \chi^d_{SO(2),\{id\}} \neq 0.$$

For further relevant topologically-minded multiparameter results, $dim\ \lambda > 1$, we refer to [Bar, Ize2, Ize&Mas&Pej&Vig, Ize&Mas&Vig] and the references given there.

We recall from §3 and fig. 3.3 that all the above approaches have a common drawback because they treat $\lambda$ and (non-minimal) period $p$, alike, as continuous parameters. In fact the "jug-handle" example, due to Alligood & Mallet-Paret & Yorke [All&M-P&Y1] shows that a continuum which is unbounded in the sense of Alexander & Yorke [Ale&Y1] may actually look quite bounded to the innocent eye. The role of generic approximations at this stage is twofold: it is possible to prove homotopy invariance of Fuller index by generic approximation [Chow&M-P] and to recover thereby the result of Alexander & Yorke. But also the concept of "snakes" [M-P&Y1, M-P&Y2] resolves the jug-handle difficulty, and generic approximations lead to an alternative result on global Hopf bifurcation [Fie2] which cannot be recovered by topological techniques so far. As we have mentioned in §3, the reason is the following. Using generic approximations we are allowed to select continua in $\Lambda \times X$ along which the minimal period jumps. Thus generic approximations have proved to be the more flexible tool, so far.

After this excursion into a world without symmetry we return to the generic equivariant setting. Trying to imitate the jug-handle example we immediately notice a difficulty. Due to theorem 5.11, $\Theta$ changes at period doubling bifurcations. This prevents us from closing up the loop as in fig. 3.3. In the case of $\mathbb{Z}_n$-symmetry, $n < \infty$, we might attempt to construct a "$j$-jug-handle" as follows. Let $m := |H/K|$ be odd and define $j := ord^*_m(2) > 1$ to be the multiplicative order of 2 $(mod\ m)$. Then the $j$-jug-handle should have the bifurcation diagram of fig. 9.1, where a flip doubling is assumed at each of the $j$ bifurcation points. Note that after $j$ flip doublings we arrive at

$$\tilde{h} \quad := \quad (k'm+2)^j\, h \qquad\qquad (mod\ \mathbb{Z})$$

as a generator for the symmetry on the secondary branch, and we may replace $\tilde{h}$ by $h$ because

$$(k'm+2)^j \quad \equiv \quad 2^j \quad \equiv \quad 1 \qquad\qquad (mod\ m)$$

and $K = \langle mh \rangle$. We did not attempt to make this example explicit. In particular, we do not know the minimal dimension of $x$ for which it may occur.

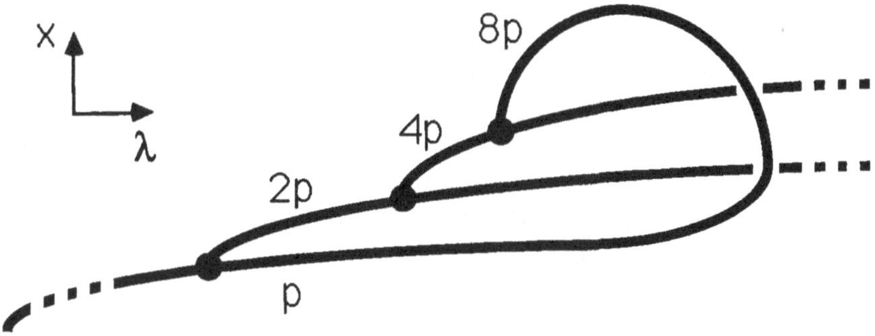


**Fig. 9.1** A $j$-jug-handle, $m = 7, j = 3$.

Let us compare the $j$-jug-handle to the jug-handle. Differently from the jug-handle, our $j$-jug-handle has not only 1 but $j$ "loose ends". And again each loose end has equivariant orbit index $\Phi = 0$, necessarily.

This phenomenon of several loose ends is intimately related to bifurcation from centers with multiple eigenvalues. In principle, each of the three loose ends of fig. 9.1 might connect to the same (nongeneric) center relating to several distinct representations of $H/K$ with nonzero crossing numbers of the purely imaginary eigenvalues. In the terminology of Alexander & Auchmuty [Ale&Au2] such a continuum would then be called global, due to the $j$-jug-handle loop. But the $\mathcal{H}$-index tells us, that some other unbounded branch must exist. In fact, following snakes, i.e. paths with $\Phi \neq 0$, we can never enter into the $j$-jug-handle loop because all loose ends have index $\Phi = 0$. Still, the absence of closed loops with just one loose end might indicate a synthesis: a global bifurcation result which keeps (non-minimal) periods continuous but does not consider a $j$-jug-handle as global. A similar refinement would be desirable for rotating waves. Who knows?

From §8 we recall another argument for controlling minimal period rather than just "period". In §§8.2, 8.3.1, we have seen how the results of Alexander & Auchmuty [Ale&Au1, Ale&Au2] cannot control symmetry because they do not control minimal period.

The remarks above should have spread sufficient doubts on whether our orbit index $\Phi$ is the "right and only" choice. Suppose we wanted to recover the results of Alexander & Auchmuty [Ale&Au1, Ale&Au2]. Then we should choose for $\Phi$ some equivariant analogue of the Fuller index. But we are still lacking such an analogue, in general.

Or we might be interested in a breaking of the orbital symmetry $H$ at flip or flop pitchforks. Then we should design an index which pushes us off the primary branch there, instead of forcing us to stay on it as our orbit index $\Phi$ does. Or we might be interested in starting from centers with $H < G$ in the generic situation: again the pitchforks spring to life. Going beyond our $\mathbb{Z}_n$ framework ($n \leq \infty$), we might ask for a suitable orbit index in generic $D_n$- or $O(2)$-equivariant bifurcation diagrams. Each task seems to require a different index, regardless whether it is approached topologically or via local singularities.

## 9.5 Homotopy invariance

Any useful index ought to be homotopy invariant. Understanding homotopy invariance of an index renders it more computable, more applicable. Here is an open question.

What are the obstructions, if any, to homotopy invariance of the global equivariant Hopf index $\mathcal{H}$ ? (9.7)

As we will explain below, we are deliberately vague here about whether to take the generic Hopf index $\mathcal{H}_n^d$ (definition 2.5) or its nongeneric counterpart $\mathcal{H}_{H_0,K_0}^{\pm d}$ (definition 2.8).

We rephrase question (9.7). The index $\mathcal{H}$ is assigned to a one-parameter family of vector fields $f(\lambda, x)$. A homotopy $f_\tau(\lambda, x)$ of such vector fields can be viewed as a two-parameter family

$$f(\lambda, \tau, x)$$

of vector fields. Let us consider any closed loop (Jordan curve)

$$c \subseteq \mathbb{R}^2$$

in parameter space $(\lambda, \tau)$. Restricting the two-parameter family $f$ to parameters in $c$ yields a one-parameter problem with associated Hopf index $\mathcal{H} = \mathcal{H}(c)$ depending on the curve $c$. Then homotopy invariance of $\mathcal{H}$ holds if, and only if,

$$\mathcal{H}(c) = 0 \qquad (9.8)$$

for any loop $c$. Contracting the loop $c$ to almost a single point, this becomes a question on singularities of equivariant two-parameter vector fields.

We relate question (9.7) to our nondegeneracy assumption (2.20.$a$), which requires stationary solutions to have only nonzero eigenvalues. If assumption (2.20.$a$) holds along the loop $c$, then necessarily

$$\mathcal{H}_{H_0,K_0}^{\pm d}(c) = 0. \qquad (9.9)$$

Indeed, on any representation space $X_r^{K_0}$ (cf. (2.22)) the number of pairs of eigenvalues crossing the imaginary axis from left to right equal those crossing back in the opposite direction, as we trace out the loop $c$ through one full cycle. This indicates that any possible obstruction to homotopy invariance of $\mathcal{H}$ is related to the occurence of zero eigenvalues. Note however that, even for nonzero eigenvalues, (9.8) need not hold if stated for the generic equivariant index $\mathcal{H}_n^d$. For example let $n = 7, d = \{1,2,4\}$ and choose a $\mathbb{Z}_n$-equivariant two-parameter vector field

$$\dot{z} \quad = \quad f(\varepsilon, z) \quad = \quad \varepsilon z + h.o.t.(z) \tag{9.10}$$

where $\varepsilon = \lambda + i\tau$, $z = x_1 + ix_2 \in \mathbb{C} \cong \mathbb{R}^2$. As usual, $\mathbb{Z}_n$ acts on $z$ by counterclockwise rotation. The $\mathbb{Z}_n$-invariant centers are given by $(\varepsilon, z) = (0+i\tau, 0)$, $\tau \neq 0$, with representation

$$r \quad = \quad \begin{cases} +1 & \text{for } \tau > 0\,, \\ -1 & \text{for } \tau < 0\,, \end{cases}$$

cf. (2.21). Choosing for $c$ a circle in the $(\lambda, \tau)$-plane centered at 0, we conclude from definition 2.5 that

$$\mathcal{H}_n^d(c) \quad = \quad -1 \quad \neq \quad 0\,, \tag{9.11}$$

because $1 \in d = \{1,2,4\}$ while $-1 \notin d$. We emphasize that nondegeneracy assumption (2.20.$a$) holds along $c$. Consistently, $\mathcal{H}_n^{-d}(c) = +1$ and

$$\mathcal{H}_{\mathbb{Z}_n,\{0\}}^{\pm d}(c) \quad = \quad -1+1 \quad = \quad 0\,.$$

It is the generic Hopf index $\mathcal{H}_n^d(c)$ which is not homotopy invariant, here. Note that a geometrically double eigenvalue 0 occurs for the linearization $D_z f(0,0)$ at $\varepsilon = 0$.

In the nonsymmetric case $\Gamma = \{id\}$, homotopy (non-)invariance of the index $\mathcal{H}$ is understood, see [Fie3]. We briefly summarize the result. The only obstruction to homotopy invariance of $\mathcal{H}$ in generic two-parameter families are **B-points**: stationary solutions with an algebraically double and geometrically simple eigenvalue 0. This singularity was analysed locally by Arnold and Bogdanov [Arn1, Arn3, Bog1, Bog2]; the normal form was obtained independently by Takens [Tak]. In [Fie3] an index $B = \pm 1$ is assigned to each $B$-point such that

$$\mathcal{H}(c) \quad = \quad \sum_{\mathrm{int}(c)} B\,, \tag{9.12}$$

where the sum ranges over all $B$-points with parameters $(\lambda, \tau)$ inside the loop $c$. Note the similarity between (9.12) and (9.11) if $c$ contains only one $B$-point.

The significance of $B$-points, and of the open question (9.7), reaches far beyond the somewhat academic topic of homotopy invariance of $\mathcal{H}$. To illustrate this let $(\lambda_0, \tau_0, x_0)$ be an isolated $B$-point, of index $+1$, $c$ a small circle around $(\lambda_0, \tau_0)$ in parameter space, and suppose the phase space $X$ is a compact manifold. Then

$$\mathcal{H}(c) \quad = \quad +1 \quad \neq \quad 0$$

regardless of how much we enlarge the loop $c$, as long as no other $B$-point leaves or enters the interior of $c$. In particular, global Hopf bifurcation occurs on $c$ (more precisely: on $c \times X$). Because $c$ and $X$ are both bounded, this implies bifurcation of a continuum of periodic solutions with arbitrarily large (virtual) periods. Mostly, we expect that virtual period = minimal period. But how can minimal periods become unbounded? A first mechanism is termination at a homoclinic solution ("blue sky catastrophe"). As the parameters vary, a periodic trajectory approaches a stationary point with some part of it, becomes a trajectory which converges to this point in both time directions (homoclinic), and disappears. This happens already near $B$-points, see [Arn1, Bog1, Bog2] and also [Arn3, Chow&Ha, Gu&Ho]. Other mechanisms of minimal periods becoming unbounded include period doubling cascades, flow plugs, etc.. An example involving Schweitzer's $C^1$-counterexample to the so-called "Seifert conjecture" is given in [Ale&Y1]. For these topics see also [Harr&Y, Y&All]. So we get continuation of homoclinics, in the first case, and necessarily complicated (chaotic?) motions in the other cases. Thinking of Šil'nikov's results on horseshoes near certain homoclinic orbits [Šil], see also [Gu&Ho], and of the "homoclinic explosion" in the $\mathbb{Z}_2$-symmetric Lorenz system, see e.g. [Spa], we can argue that continuation of homoclinics may likewise lead to chaotic motions. Especially in equivariant systems, continuation of homoclinics should yield interesting results as the Lorenz model suggests.

Returning to $\mathbb{Z}_n$-equivariant systems we find analogues of $B$-points. Such points have been studied by Arnold [Arn2, Arn3] and, in case $n = 2, 3$, by Horozov [Hor]. The case $n = 4$ is not resolved completely. The normal forms begin as in (9.10). The normal form for "$B$-points" with $D_n$- or $O(2)$-symmetry can be obtained, to any order, from the results of Cushman & Sanders [Cu&San]. The dynamics of the $O(2)$-case are discussed extensively in [Da1, Da2, Da&Arm, Da&Kno, Gu]. For a discussion of other bifurcations in codimension two we refer to [Chos&Dem&Io, Chos&Go&Key, Go&St2]. Although the results of Arnold and Horozov [Arn2, Hor] were originally motivated by resonant bifurcation of tori, we expect them to play a central role for global equivariant Hopf bifurcation in two parameters and for understanding homotopy invariance of the index $\mathcal{H}$.

## 9.6 Manifolds of solutions

In nondegeneracy assumption (2.20.$a$) we have required the branches $(\lambda, x(\lambda)) \in \Lambda \times X^{H_0}$ to have only nonzero eigenvalues. One might wonder whether the action of the compact Lie group $\Gamma$ forces degeneracies which make assumption (2.20.$a$) self-contradictory. Let $x_0$ be a stationary solution and $\Gamma_{x_0} = H_0$ its isotropy. Then, acting on $x_0$ by $\Gamma$, we obtain a solution manifold diffeomorphic to the coset space $\Gamma/\Gamma_{x_0}$. Restricting our attention to $X^{H_0}$, however, we are interested in this manifold only as far as it intersects $X^{H_0}$. This part is diffeomorphic to $W(\Gamma_{x_0}) := N(\Gamma_{x_0})/\Gamma_{x_0}$, where $N(\Gamma_{x_0})$ and $W(\Gamma_{x_0})$ denote the normalizer and the "Weyl group" of $\Gamma_{x_0}$, respectively. Unless $W(\Gamma_{x_0})$ is discrete, the (group-) orbit of $x_0$ in $X^{H_0}$ forces degeneracies.

Frozen waves are an example. Indeed, let $\Gamma = SO(2) = \mathbb{R}/\mathbb{Z}$, $\Gamma_{x_0} = H_0 = \mathbb{Z}_n$ , $n < \infty$. Then $N(\Gamma_{x_0}) = SO(2)$, $W(\Gamma_{x_0}) \cong SO(2) = S^1$ and we obtain a circle of stationary solutions, a frozen wave (cf. definition 1.1; $H_0$ was called $K$ there). By theorem 5.11, frozen waves occur only for isolated parameter values in generic one-parameter systems. If

$$dim\, W(\Gamma_{x_0}) \quad \geq \quad 2\,,$$

we obtain a higher-dimensional manifold. But we expect the flow on this manifold to be nonstationary for generic $\Gamma$-equivariant one-parameter vector fields. Then frozen waves are the typical case of manifolds of stationary solutions.

Still, we may ask for bifurcation of solution manifolds, given by the action of $W(\Gamma_{x_0})$, and for the flow on such invariant manifolds. In [Go&St1, §12], Golubitsky & Stewart ask for a natural bifurcation context involving $W(\Gamma_{x_0}) = SU(2) \cong S^3$. They observe that such a bifurcation would require three additional parameters, besides $\lambda$, for its description. Here is an example. Let $X := \mathbb{R}^4 \cong \mathbb{H}$ be the space of quaternions, and let $SU(2) \cong S^3$, the group of unit quaternions, act by multiplication on $X$. Consider an $SU(2)$ equivariant vector field

$$f(\lambda,\cdot): \quad X \rightarrow X.$$

Then $f(\lambda,0)=0$. Writing $\dot{x}=f(\lambda,x)$ in polar coordinates, $r=|x|$, we get

$$\dot{r} = r\cdot a(\lambda,r),$$

where $a$ is independent of the angle coordinates by $SU(2)$ equivariance of $f$. Assuming $a(0,0)=0$ and the transversality condition $D_\lambda a(0,0)\neq 0$, we obtain $\lambda=\lambda(r)$, locally, such that

$$0 = r\cdot a(\lambda(r),r).$$

This constitutes a flow-invariant 3-sphere of radius $r$ for the vector field $f(\lambda(r),\cdot)$, bifurcating from the trivial solution. Identifying this invariant sphere with $SU(2)$ again, we see that $SU(2)$ acts transitively on it. In particular, the flow on it is determined precisely by $f(\lambda(r),x^*)$, where $x^*$ corresponds to $id\in SU(2)$. Because $f(\lambda(r),x^*)$ has to be tangent to the 3-sphere this leaves us with three additional free parameters which determine the flow on that sphere. At the bifurcation point $\lambda=0$, these three parameters can be computed from the linearization $D_x f(0,0)x^*$. Replacing $SU(2)$ by $SO(2)$, this is precisely analogous to Hopf bifurcation of rotating waves. The one free additional parameter there is the speed of rotation, or the minimal period.

This guides us to ask another question: what is the global "bifurcation" behavior of solution manifolds, obtained by the action of the Weyl group $W(\Gamma_{x_0})$? And what is the flow on this manifold? For $W(\Gamma_{x_0}) = SO(2)$ we have tried to answer this question in theorem 2.9. We obtain global bifurcation results for rotating/frozen waves. They resemble stationary bifurcation results, because they come from the action of the Weyl group. Factoring out this action, conceptually, we may view theorem 2.9 as a global result on bifurcation of relative equilibria for $dim\ W(\Gamma_{x_0})=1$. But this is only the simplest case, and $dim\ W(\Gamma_{x_0})\geq 2$ is entirely open to the public. However, we still do not know any natural application context for these phenomena, other than the mathematical construction given above.

In §8 we have encountered examples of a different kind of solution manifolds: invariant 2-tori (cf. §8.3.8). They were bifurcating from rotating waves. Thus the Weyl group is $SO(2)$, here, and its action foliates the 2-torus periodically. The dynamics on the torus may be quasiperiodic, but equivariant with respect to the $SO(2)$-action, of course. En passant, we note that this action makes the rotation number depend differentiably on the parameters. For rational rotation numbers, the torus consists entirely of periodic solutions. All these claims are easy to verify locally near the rotating wave from which the torus bifurcates. Just use a local Poincaré-section $S$ to the rotating wave and the Weyl group action, and project the flow on the torus onto $S$. The rotating wave then becomes a stationary point in $S$, a relative equilibrium, and the torus bifurcation is a plain and simple Hopf bifurcation in $S$. Thus these tori can be viewed as relative periodic solutions to the $SO(2)$-action.

Of course, this is a local argument. It leads to another open question:

> What can be said about global bifurcation of 2-tori, foliated by an $SO(2)$-action? (9.13)

Being unable to answer this question, now, we may happily generalize it. We may wonder about global bifurcations of relative periodic solutions, relative homoclinics, etc., for Weyl group actions with $dim\ W(\Gamma_{x_0})\geq 2$. But our discussion, which has started out at blue sky, is about to dissolve in catastrophic clouds. So we stop.

# 10 Appendix on genericity

## 10.1 Outline

In this appendix we give a condensed but complete and, unfortunately, technical account of the transversality arguments leading to the local generic results formulated in theorem 5.2, lemma 5.9, and theorem 5.11. For a general background on transversality theory see [Ab&Ro]. Due to technical complications, we will use notation more freely than in previous sections whenever misunderstandings can be excluded by the context.

The outline is the following. In §10.2 we present a transversality theorem (theorem 10.2), due to Tromba [Tro], which is basic to the whole section. This theorem can be applied in an analytic semigroup setting to yield genericity results [Fie2]. To apply theorem 10.2, we have to analyse codimensions of some varieties of equivariant matrices with prescribed eigenvalues. The basic results are collected in §10.3, proposition 10.3. The transversality machinery is then used in §10.4, illustratively, to prove theorem 5.2 on generic Hopf bifurcation. Concerning secondary bifurcations, we first consider case (c) of theorem 5.11, i.e. finite cyclic equivariance $G = \mathbb{Z}_n$ of

$$\dot{x} = f(\lambda, x)\,.$$

In lemma 10.6, we construct the required perturbations, locally, for a given periodic solution $(\lambda, x)$ with **minimal period** $p$ and **known symmetry** $(H, K, \Theta)$. To overcome this necessary restriction, we then describe a localization process in §10.6 which provides the precise technical setting for our application of transversality. These long preparations pay off in §10.7 in our proof of lemma 5.9. The proof involves an induction over increasing minimal period (as usual: [Ab&Ro, All&M-P&Y2, Fie2, Kup, Med, Peix, Smal, Sot]), and in addition over decreasing $\mathbf{h} = |H|$. We finish this appendix in §§10.8, 10.9 by sketching simplifications and necessary modifications of these arguments, which then take care of the simple assertions of theorem 5.11 in cases (a) and (b), $G = \mathbb{Z}_\infty$.

## 10.2 Abstract transversality

We start with abstract transversality in an infinite-dimensional setting. Let $\mathcal{U}, J, \mathcal{Y}$ be open subsets of Banach spaces, and

$$V = \bigcup_{i=0}^{j_0} V_j \tag{10.1}$$

a finite dimensional variety in $\mathcal{Y}$ with stratification $V_j$. More precisely, let $V$ be a closed separable subset of $\mathcal{Y}$ consisting of a finite disjoint union of (not necessarily closed) embedded $C^k$ submanifolds $V_j$ of $\mathcal{Y}$ (called strata of $V$), with decreasing finite dimensions $\mathit{dim}\ V_j$, $\mathit{dim}\ V := \max_j \mathit{dim}\ V_j = \mathit{dim}\ V_0$, such that $V_{j_1} \cup V_{j_1+1} \cup \ldots \cup V_{j_0}$ is closed for any $j_1 \le j_0$. Let

$$\begin{aligned} R: \quad \mathcal{U} \times J &\to \mathcal{Y} \\ (f, w) &\mapsto R(f, w) \end{aligned} \tag{10.2}$$

be a $C^k$-map, $k \ge 1$. We collect some terminology.

**10.1 Definition :**

*We call $R$* **Fredholm**, *if $R$ is Fredholm with respect to $w \in J$. That is, we assume that the image $im(D_w R)$ is closed with finite dimensional complement and that $D_w R$ has finite* **Fredholm index**

$$i(R(f,w)) \quad := \quad nul\,(D_w R(f,w)) \; - \; def\,(D_w R(f,w))$$

*for any $f \in \mathcal{U}, w \in J$ (cf. [Kato]). Here $nul$, $def$ denote dimension of the kernel, codimension of the image, respectively.*

*We call $R$* **proper** *with respect to $V$ on a subset $\mathcal{K}$ of $J$, if $R(f_n, w_n) \in V, f_n \in \mathcal{U}, w_n \in \mathcal{K}, f_n \to f \in \mathcal{U}$ together imply that some subsequence of $w_n$ converges to a limit in $\mathcal{K}$.*

*We call $R$* **transverse** *to $V$ at $(f,w)$ (in symbols: $R \pitchfork_{(f,w)} V$) if $R(f,w) \in V_j$ for some $j$ implies that $im\ DR(f,w)$ together with the tangent space $T_{R(f,w)}V_j$ spans the Banach space of $\mathcal{Y}$, i.e.*

$$T_{R(f,w)}V_j \; + \; im\ DR(f,w) \quad = \quad T_{R(f,w)}\mathcal{Y}\,.$$

*In particular $R \pitchfork_{(f,w)} V$, if $R(f,w) \notin V$. The related notations $R(f,\cdot) \pitchfork_w V$, $R \pitchfork V$ etc. are self-explanatory.*

The following theorem is essentially due to Tromba [Tro] and the modification which we use here is contained in [Fie2, §2]. For a finite dimensional background we recommend [Ab&Ro], again.

**10.2 Theorem :**

*Let $\mathcal{U}, J, \mathcal{Y}, R, V$ be as in (10.1), (10.2) above. In addition, we assume for the $C^k$-map $R$ that*

$$R \text{ is Fredholm on } R^{-1}(V) \text{ with constant index } i = i(D_w R(f,w)) \in \mathbb{Z}; \tag{10.3}$$

$$R \text{ is proper on a subset } \mathcal{K} \text{ of } J; \tag{10.4}$$

$$R \pitchfork V\,; \tag{10.5}$$

$$k \; > \; \max\{0, i + dim\ V\}\,. \tag{10.6}$$

*Then the set*

$$\mathcal{G}_V \quad := \quad \{f \in \mathcal{U} \;\; | \;\; R(f,\cdot) \pitchfork_w V \text{ for all } w \in \mathcal{K}\} \tag{10.7}$$

*is generic in $\mathcal{U}$.*

*If we assume moreover that*

$$V \text{ is a manifold, or} \tag{10.8.a}$$

$$i + dim\ V \quad < \quad 0\,, \tag{10.8.b}$$

*then $\mathcal{G}_V$ is open and dense in $\mathcal{U}$.*

For a proof see [Fie2, theorem 2.5, corollary 2.6]. The basic idea is to consider manifolds $V$ and the projection to the $f$-component

$$\pi: \qquad R^{-1}(V) \quad \to \quad \mathcal{U}\,.$$

By (10.5) and the implicit function theorem, $R^{-1}(V)$ is a manifold. Moreover, $\mathcal{G}_V$ consists precisely of the regular values of $\pi$. Because $\pi$ turns out to be proper and Fredholm with the same index as $R$, invoking the Smale-Sard theorem [Sma2] completes the proof. We omit all further details.

From the definitions it is obvious that

$$\mathcal{G}_V \;=\; \{f \in \mathcal{U} \mid R(f,\mathcal{K}) \cap V = \emptyset\},$$

if assumption (10.8.$b$), $i + \dim V < 0$, holds.

## 10.3 Varieties of equivariant matrices

In a typical application of theorem 10.2, $f$ will be the nonlinearity in

$$\dot{x} \;=\; f(\lambda, x). \tag{1.1}$$

We will define

$$R \;=\; (\Pi_0, D_x\Pi_0), \qquad w = (\lambda, x),$$

for a $K$-equivariant "Poincaré" map $\Pi_0$ as in (5.12), and then put $V = \{0\} \times \mathcal{M}_j^K$ for some appropriate set $\mathcal{M}_j^K$ of $K$-equivariant matrices. To this end, we now define some useful sets of such matrices and calculate their (co-)dimensions. "Useful" matrices are those which pertain to our conditions on flips, flops, flip-flops and generic centers as given in §§3 and 5. To fix notations let $K = \mathbb{Z}_k$, $k \le \infty$ with real representations $\rho^+$, and possibly $\rho^-$, and with complex representations $\rho_r$ as irreducible types (cf. §§2,3). Let $K$ act on $\mathbb{R}^N$ (not necessarily irreducibly) by $\rho$. Writing $spec(M)$ for the spectrum of a matrix $M$ and $mult(\mu_0)$ for the algebraic multiplicity of $\mu_0$ ($mult := 0$ outside $spec(M)$), we define the sets

$\mathcal{M}(\mathbb{K}, N)$: all $N \times N$ matrices over $\mathbb{K} = \mathbb{R}$ or $\mathbb{C}$; (10.9)

$\mathcal{M}^K(\mathbb{R}, N)$: all $K$-equivariant matrices $M \in \mathcal{M}(\mathbb{R}, N)$;

$\mathcal{M}(\mu_0, m)$: all $M \in \mathcal{M}(\mathbb{R}, N)$with $mult(\mu_0) \ge m$;

$\mathcal{M}^K(\mu_0, m)$: $=\; \mathcal{M}(\mu_0, m) \cap \mathcal{M}^K(\mathbb{R}, N)$;

$\mathcal{M}_{1,0}$: all $M \in \mathcal{M}^K(\mathbb{R}, N)$ such that $spec(M)$ contains an eigenvalue $\mu_0$ with $Re\,\mu_0 = 0$;

$\mathcal{M}_{2,0}$: all $M \in \mathcal{M}_{1,0}$ such that
$mult(0) \ge 2$, or (10.10.$a$)
the sum of multiplicities of $\mu_0 \in spec(M)$ with $Re\,\mu_0 = 0$ exceeds 2; (10.10.$b$)

$\mathcal{M}'_{2,0}(k)$: all $M \in \mathcal{M}^K(\mathbb{R}, N)$ such that $spec(M)$ contains an eigenvalue $\mu_0 = 2\pi i \cdot e/k$ for some $e \in \mathbb{Z} \setminus \{0\}$ and fixed integer $k$.

$\mathcal{M}_{1,1}$: all $M \in \mathcal{M}^K(\mathbb{R}, N)$ such that $spec(M)$ contains an eigenvalue $\mu_0$ with $|\mu_0| = 1$;

$\mathcal{M}_{2,1}$ : all $M \in \mathcal{M}_{1,1}$ such that
$mult(+1) + mult(-1) \geq 2$, or (10.11.a)
the sum of multiplicities of $\mu_0 \in spec(M)$ with (10.11.b)
$|\mu_0| = 1$ exceeds 2;

$\mathcal{M}'_{2,1}(k)$ : all $M \in \mathcal{M}^K(\mathbb{R}, N)$ such that $spec(M)$ contains a primitive $e$-th root of unity for some $3 \leq e \leq k$ and fixed $k$.

Note that these sets define algebraic varieties in the linear vector space $\mathcal{M}^K(\mathbb{R}, N)$. We calculate real codimensions in that space.

**10.3 Proposition :**

$$codim\ \mathcal{M}^K(\mu_0, m) = \begin{cases} m & if\ \mu_0 \in \mathbb{R}\,, \\ 2m & if\ \mu_0 \notin \mathbb{R}\,. \end{cases} \tag{10.12.a}$$

$$codim \bigcap_l \mathcal{M}^K(\mu_l, m_l) = \sum_l codim\ \mathcal{M}^K(\mu_l, m_l) \tag{10.12.b}$$

*if the pairs $\{\mu_l, \bar{\mu}_l\}$ are mutually distinct; and*

$$codim\ \mathcal{M}_{j,\beta} = j, \qquad codim\ \mathcal{M}'_{j,\beta} = j \tag{10.12.c}$$

*whenever $\mathcal{M}_{j,\beta}$ or $\mathcal{M}'_{j,\beta}$ is defined.*

For $K = \{0\}$, the proof could follow essentially Brunovský [Bru]. For the convenience of the reader we give our own account, including nontrivial $K$. Basically, the proof of proposition 10.3 reduces to an application of results by Arnold and Galin on codimensions of sets of matrices, cf. [Arn3, §30]. The next lemma takes care of $K$-equivariance. We recall that $K$ is said to act isotypically on $\mathbb{R}^N$ if only one type of all the inequivalent irreducible representations of $K$ actually occurs.

**10.4 Lemma :**
*Let $K$ act isotypically on $\mathbb{R}^N$ by a real or complex representation $\rho$. Then we have isomorphisms*

$$\mathcal{M}^K(\mathbb{R}, N) \cong \mathcal{M}(\mathbb{R}, N) \qquad \text{for } \rho \text{ real}, \tag{10.13.a}$$

$$\mathcal{M}^K(\mathbb{R}, N) \cong \mathcal{M}(\mathbb{C}, N/2) \qquad \text{for } \rho \text{ complex}. \tag{10.13.b}$$

**Proof of lemma 10.4 :**
If $\rho = \rho^+$ or $\rho^-$ is real, then $K$-equivariance is automatic by linearity. Therefore (10.13.a) is trivial.

Suppose now that $\rho = \rho_r$ is complex. We decompose $\mathbb{R}^N$ into irreducible subspaces, each of which we identify with $\mathbb{C}$ such that

$$\rho(h) = e^{2\pi i r h}, \qquad \text{for all } h \in K \leq \mathbb{R}/\mathbb{Z}\,,$$

acts by complex multiplication on each subspace. This defines an isomorphism $\iota$ from $\mathbb{R}^N$ to $\mathbb{C}^{N/2}$ such that the following diagram commutes

$$\begin{array}{ccc} \mathbb{R}^N & \xrightarrow[\rho(h)]{} & \mathbb{R}^N \\ \iota\downarrow & & \iota\downarrow \\ \mathbb{C}^{N/2} & \xrightarrow[exp\,(2\pi i r h)]{} & \mathbb{C}^{N/2} \end{array}$$

We define a map

$$\begin{array}{rcl} \mathcal{M}^K(\mathbb{R},N) & \rightarrow & \mathcal{M}(\mathbb{C},N/2) \\ M & \mapsto & \hat{M} := \iota M \iota^{-1}. \end{array}$$

Pick some $h$ such that $exp\,(i\vartheta) := exp\,(2\pi i r h)$ is not real. Then

$$\begin{array}{rcll} \rho(h^{\pm 1})M & = & M\rho(h^{\pm 1}) & \Longleftrightarrow \\ e^{\pm i\vartheta}\hat{M} & = & \hat{M}e^{\pm i\vartheta}. & \end{array}$$

Thus $\hat{M}$ is complex linear if and only if $M$ is $K$-equivariant. Therefore $\mathcal{M}^K(\mathbb{R},N)$ and $\mathcal{M}(\mathbb{C},N/2)$ are isomorphic, by $M \mapsto \hat{M}$, and the proof is complete. □

**Proof of proposition 10.3 :**

To prove (10.12.$a$), we may restrict our attention to the case

$$m = \begin{cases} N & \text{if } \mu_0 \in \mathbb{R} \\ N/2 & \text{if } \mu_0 \notin \mathbb{R}. \end{cases}$$

Indeed, we can use the equivariant eigenprojection associated to $\mu_0$ to achieve this reduction.

Similarly, we may assume that $K$ acts isotypically on $\mathbb{R}^N$. Otherwise we decompose $\mathbb{R}^N$ into representation subspaces, then calculate codimensions for each restricted matrix separately, and finally add up these codimensions.

The remaining parts of the proof rely crucially on Arnold's results on versal deformations of matrices [Arn3, §30]. These results are based on the Arnold normal form of matrices; see [Arn3, §30, theorem 2]. Suppose first that the action $\rho$ of $K$ on $\mathbb{R}^N$ is real, $\rho = \rho^+$ or $\rho^-$. By lemma 10.4 it is then sufficient to prove

$$codim_R\ \mathcal{M}(\mu_0, m) \quad = \quad N, \tag{10.14}$$

where $codim_R$ is understood in $\mathcal{M}(\mathbb{R},N)$. From $m = N$ and [Arn3, §30, theorem 1], assertion (10.14) is immediate.

Next suppose that $\rho$ is of complex type. By lemma 10.4 it is then sufficient to prove

$$codim_c\ \mathcal{M}(\mu_0, N/2) \quad = \quad N/2\,, \tag{10.14$'$}$$

where $codim_c$ is understood as complex codimension in $\mathcal{M}(\mathbb{C},N/2)$, this time. Indeed, (10.14)$'$ implies

$$codim_R\ \mathcal{M}^K(\mu_0, m) \quad = \quad N,$$

because real codimension in $\mathcal{M}^K(\mathbb{R},N)$ equals twice the complex codimension in $\mathcal{M}(\mathbb{C},N/2)$. Note here that the multiplicity of $\mu_0$ for the complexification in $\mathcal{M}(\mathbb{C},N/2)$ becomes $N/2$, regardless of whether $\mu_0$ is real or complex.

Again, $(10.14)'$ is immediate from [Arn3, §30]. This proves (10.12.*a*).

To prove (10.12.*b*) we just decompose any matrix near $\bigcap \mathcal{M}^K(\mu_l, m_l)$ into blocks by the eigenprojections associated to $\{\mu_l, \bar{\mu}_l\}$. Then the respective codimensions add up, proving (10.12.*b*).

To prove (10.12.*c*), we consider each set $\mathcal{M}_{j,\beta}$ individually. Consider $\mathcal{M}_{1,0}$ first. Certainly *codim* $\mathcal{M}_{1,0} > 0$. Let $M_0 \in \mathcal{M}_{1,0}$ have only one simple pair of purely imaginary eigenvalues $\mu_0, \bar{\mu}_0 \neq 0$. Perturbing $\mu_0$ to $\mu_0 + \lambda$, $\lambda \in \mathbb{R}$, we see that *codim* $\mathcal{M}_{1,0} = 1$, which is the codimension of the stratum of $M_0$. This is a trivial consequence of putting $M_0$ into Arnold normal form. Because $\mu_0$ is simple it can also be seen directly, using the eigenprojection associated to the local continuation of $\mu_0$. By exponentiation, the codimension of $\mathcal{M}_{1,1}$ likewise equals 1.

Consider $\mathcal{M}'_{2,0}(k)$, $\mathcal{M}'_{2,1}(k)$ next. Because these sets are defined via eigenvalues $\mu_0 \notin \mathbb{R}$, their codimensions are 2 by (10.12.*a*).

The set $\mathcal{M}_{2,0}$ is the union of two sets given by (10.10.*a*) respectively (10.10.*b*). The set given by (10.10.*a*) has codimension 2 in $\mathcal{M}^K(\mathbb{R}, N)$, by (10.12.*a*). Next suppose that (10.10.*b*) holds. The set

$$\mathcal{M}_{3,0} := \{M \in \mathcal{M}_{2,0} \mid mult(\mu_0) \geq 2 \quad \text{for some } \mu_0 \text{ with } Re\,\mu_0 = 0 \neq Im\,\mu_0\}$$

has $codim_{\mathbb{R}} \geq 3$ in $\mathcal{M}^K(\mathbb{R}, N)$. Indeed, the complex Jordan decomposition of $M \in \mathcal{M}_{3,0}$ contains

$$\begin{pmatrix} \mu_0 & & & \\ 1 & \ddots & & \\ & \ddots & \ddots & \\ & & 1 & \mu_0 \end{pmatrix} \quad \text{or} \quad \begin{pmatrix} \mu_0 & 0 \\ 0 & \mu_0 \end{pmatrix}.$$

As in [Arn3, fig. 109(b)], complex perturbations transverse to $\mathcal{M}_{3,0}$ are given explicitly in Arnold normal form by

$$\begin{pmatrix} \lambda & & & \\ \lambda_2 & 0 & & \\ \vdots & & \ddots & \\ \lambda_m & & & 0 \end{pmatrix} \quad \text{or} \quad \begin{pmatrix} \lambda & 0 \\ \lambda_2 & \lambda' \end{pmatrix},$$

with $\lambda, \lambda' \in \mathbb{R}$, $\lambda_2, \ldots, \lambda_m \in \mathbb{C} \cong \mathbb{R}^2$, $m \geq 2$. This yields $codim_{\mathbb{R}} \geq 3$ respectively $\geq 4$. All elements of $\mathcal{M}_{2,0} \setminus \mathcal{M}_{3,0}$ yield codimensions $\geq 2$, by (10.12.*a*, *b*). This proves *codim* $\mathcal{M}_{2,0} = 2$.

The case of $\mathcal{M}_{2,1}$ is similar. Note that (10.11.*b*) is related to (10.10.*b*) by exponentiation, yielding equal codimensions. Applying (10.12.*a*, *b*) the codimension associated to $\mathcal{M}_{2,1}$ is also 2, and the proof of proposition 10.3 is complete. □

## 10.4 Generic centers

With the basic transversality machinery at hand we now turn to genericity. We recall our notion of genericity from §1.5, definition 1.3. Given some property $\mathcal{E}$ of stationary or periodic solutions of

$$\dot{x} = f(\lambda, x), \tag{1.1}$$

we call $\mathcal{E}$ generic, if the subset

$$\mathcal{G} \quad := \quad \{f \in \mathcal{F} \mid \mathcal{E} \text{ holds}\}$$

is generic in $\mathcal{F}$. Recall from (1.36) that $\mathcal{F}$ is the set of $G$-equivariant maps in $C^4(\Lambda \times X, X)$, putting $H_0 = G = \mathbb{Z}_n$ and $K_0 = \{0\}$ in (1.36).

Genericity can be "localized" as follows. Let

$$X \quad = \quad \bigcup_j \mathcal{B}_j$$

be a countable covering of $X$ with closed balls $\mathcal{B}_j$, and define

$$\mathcal{G}(\mathcal{B}_j) \quad := \quad \{f \in \mathcal{F} \mid \mathcal{E} \text{ holds for all } (\lambda, x) \in \mathcal{B}_j\}. \tag{10.15}$$

If $\mathcal{G}(\mathcal{B}_j)$ is generic, for each $j$, then the countable intersection

$$\mathcal{G} \quad = \quad \bigcap_j \mathcal{G}(\mathcal{B}_j)$$

is still generic. Thus it will be sufficient to prove that each $\mathcal{G}(\mathcal{B}_j)$ is generic. This localizes the proofs because compactness becomes effective on $\mathcal{B}_j$. Also, to investigate $\mathcal{G}(\mathcal{B}_j)$ it suffices to know $f$ on a slightly larger open ball $\mathcal{B}'_j$, and we may use any cut-off procedure for $f$ beyond $\mathcal{B}'_j$. We choose not to mention $\mathcal{B}_j, \mathcal{B}'_j$ henceforth, omitting the index $j$, and replacing $\mathcal{G}$ by $\mathcal{G}(\mathcal{B})$ and $\mathcal{F}$ by the Banach space

$$\mathcal{F} \quad = \quad \{f \in C^4(\mathcal{B}', \mathbb{R}^N) \mid f(\lambda, \gamma x) \; = \; \gamma f(\lambda, x) \text{ for all } \gamma \in G\} \tag{10.16}$$

tacitly, to avoid notational complications.

As a first illustration of the machinery developed so far we prove theorem 5.2 on generic $H$-centers.

**Proof of theorem 5.2 :**

We have to prove that, generically, $H$-centers are generic for any $H \leq G$, cf. definition 2.2. Fixing $H$, we need consider only $H$-centers $(\lambda, x)$ with isotropy $G_x = H$.

To apply transversality theorem 10.2 let

$$\mathcal{U} := \mathcal{F}, \qquad \mathcal{Y} := X^H \times \mathcal{M}^H(\mathbb{R}, N).$$

Here $\mathcal{M}^H$ is defined as $\mathcal{M}^K$, with $H$ replacing $K$, of course. Let

$$\mathcal{J} \quad := \quad \{(\lambda, x) \in (\Lambda \times X^H) \cap \mathcal{B}' \mid G_x = H\}.$$

Note that $\mathcal{J}$ is in fact open in $\mathcal{Y}$, and isotropy on $\mathcal{J}$ is given exactly by $H$. For $\mathcal{K}$ we take any closed subset

$$\mathcal{K} \quad \subseteq \quad \mathcal{J} \cap \mathcal{B}'.$$

Finally we define

$$R(f, w) := (f(w), D_x f(w)), \qquad w := (\lambda, x),$$

and $V \subseteq \mathcal{Y}$ shall have the form

$$V := \{0\} \times V'$$

with subvarieties $V'$ of $\mathcal{M}^H(\mathbb{R}, N)$ to be specified below.

Then all assumptions of theorem 10.2 are satisfied. Indeed,

$$R: \quad \mathcal{F} \times J \quad \to \quad \mathcal{Y}$$

is $C^3$, Fredholm of index

$$i \quad = \quad dim\, \Lambda \; - \; dim\; \mathcal{M}^H(\mathbb{R}, N) \quad = \quad 1 \; - \; dim\; \mathcal{M}^H(\mathbb{R}, N),$$

proper on $\mathcal{K}$, and transverse to anything because the linearization

$$\begin{array}{ccccc} g & \mapsto & D_f R(f, \lambda, x) \cdot g & = & (g(\lambda, x),\; D_x g(\lambda, x)) \\ \mathcal{F} & \to & X^H \times \mathcal{M}^H(\mathbb{R}, N) & & \end{array}$$

is surjective. Here we use that the isotropy $G_x$ of $x$ is just $H$. In case $G_x > H$, surjectivity would contradict the $G$-equivariance of $g \in \mathcal{F}$. Finally, differentiability condition (10.6) is also satisfied:

$$\begin{aligned} k &= 3 > 1 \geq 1 - codim\, V' = 1 - (dim\; \mathcal{M}^H(\mathbb{R}, N) - dim\, V) \\ &= i + dim\, V\,. \end{aligned}$$

Applying theorem 10.2, we first take $V' = \mathcal{M}^H(\mathbb{R}, N)$. Then $\mathcal{G}_V$ is open and dense, by (10.8.$a$). Moreover, $f \in \mathcal{G}_V$ iff 0 is a regular value of $f$. In that case $f^{-1}(0)$ is a one-dimensional embedded submanifold, locally near any $w \in \mathcal{K}$.

Next we redefine $\mathcal{U} := \mathcal{G}_V$ and afterwards we pick a new $V' := \mathcal{M}_{2,0}$. Then (10.8.$b$) holds by proposition 10.3:

$$i + dim\, V \quad = \quad 1 - codim\, V' \quad = \quad -1 \quad < \quad 0\,.$$

Hence the new $\mathcal{G}_V$ is again open and dense, and

$$f(\lambda, x) = 0 \quad \Rightarrow \quad D_x f(\lambda, x) \notin \mathcal{M}_{2,0} \tag{10.17}$$

for $f \in \mathcal{G}_V$, $(\lambda, x) \in \mathcal{K}$. Now suppose that $(\lambda, x) \in \mathcal{K}$ is an $H$-center. Then $D_x f(\lambda, x)$ does possess some purely imaginary nonzero eigenvalues. By (10.17) they are simple, and 0 is not an eigenvalue of $D_x f(\lambda, x)$. This proves condition (2.2.$a$) for a generic center.

Replacing $\mathcal{U}$ by $\mathcal{G}_V$, once more, we may pick a new $V' := \mathcal{M}_{1,0}$. Because $R$ hits $V = \{0\} \times V'$ only in the stratum of codimension 1 given by

$$\mathcal{M}_{1,0} \setminus \mathcal{M}_{2,0}\,,$$

transverse crossing of the purely imaginary pair as in (2.3.$b$) follows for $f$ in the open dense set $\mathcal{G}_V$. In particular, centers are then isolated.

Finally, the curvature condition (5.3.$c$) can be satisfied by an explicit local perturbation which was given in [Fie2], equivariantly as it turns out. This completes the proof. □

We define

$$\mathcal{G}_0(\mathcal{B}): \quad \text{the set of those } f \in \mathcal{F} \text{ for which all centers in } \mathcal{B} \text{ are generic.} \tag{10.18}$$

This set is generic, by theorem 5.2. Obviously, $\mathcal{G}_0(\mathcal{B})$ is also open. By condition (2.3.$c$), the branch of periodic solutions bifurcating from a generic center consists of hyperbolic periodic solutions: locally, as we recall. Denoting the flow of (1.1) through $(\lambda, x)$ by

$$\Psi_t(f, \lambda, x)\,,$$

we now express some uniformity of this hyperbolicity property for later use.

**10.5 Lemma :**

*Let $f_0 \in \mathcal{G}_0(\mathcal{B})$ and $T_0 > 0$ be given. Then there exists an open neighborhood $\mathcal{N}_0$ of $\{(\lambda, x) \in \mathcal{B} \mid f_0(\lambda, x) = 0\}$, and an open neighborhood $\mathcal{U}_0$ of $f_0$ in $\mathcal{G}_0(\mathcal{B})$ such that the following continuity property holds.*

*If $f$ in $\mathcal{U}_0$, and if $x(t) = \Psi_t(f, \lambda, x)$ is any nonstationary periodic solution with minimal period $\leq T_0$ such that*

$$(\lambda, x(t)) \in \mathcal{N}_0, \quad \text{for all } t,$$

*then $(\lambda, x(t))$ is hyperbolic.*

**Proof :**

We argue indirectly. Suppose there exist converging sequences

$$\begin{array}{ccc} f_j & \to & f_0 \\ (\lambda_j, x_j(t)) & \to & (\lambda_0, x_0), \quad \text{uniformly in } t, \end{array}$$

with $f_0(\lambda_0, x_0) = 0$ and with the minimal period of the non-hyperbolic periodic solutions $(\lambda_j, x_j(t))$ being uniformly bounded by $T_0$. Then $(\lambda_0, x_0) \in \mathcal{B}$ is a center of $f_0$, by virtual symmetry corollary 4.6. It is a generic center, because $f_0 \in \mathcal{G}_0(\mathcal{B})$. An application of the implicit function theorem shows that this center together with the emanating local branch of hyperbolic periodic solutions persists, uniformly for small perturbations $f_j$ of $f_0$. This contradicts the assumption that $(\lambda_j, x_j(t))$ is non-hyperbolic, for large $j$. Thus the lemma is proved. □

## 10.5 Perturbations

To study genericity of secondary bifurcations of periodic solutions we consider case (c) of theorem 5.11, first. In lemma 10.7 below, we give a rather local analysis of perturbations of the "Poincaré map" $\Pi_0$, introduced in (5.12) above. Of course, these perturbations aim at the crucial transversality assumption (10.5) in theorem 10.2. Even in the absence of any equivariance, the transversality assumption can be expected to hold only for periodic solutions $\Psi_t(f, \lambda, x) = x$ with **minimal period** $t = p$. See [Ab&Ro] for a detailed discussion of this technical obstacle which is circumvented by an induction over the ranges of minimal period. See also [All&M-P&Y2, Fie2, Kup, Med, Peix, Smal, Sot]. Including equivariance will require an additional induction over decreasing $\mathbf{h} = |H|$, for each level of minimal periods. In §10.6 below, we describe a localization procedure which leads to sets $J, K$ which are suitable for an application of theorem 10.2.

Assume now that $G$ is finite, $x_0(t) = \Psi_t(f, \lambda_0, x_0)$ is periodic with minimal period $p_0$, symmetry $(H, K, \Theta)$, and $h \in H$ satisfies (5.11.*a*-*c*). Let $S = \{x_0\} + S'$, $S' = \langle \dot{x}_0(0) \rangle^\perp$ be a fixed Poincaré section. Then $\Pi_0 = \Pi_0(g, \lambda, x)$ is defined locally, for $(g, (\lambda, x))$ in a neighborhood $\mathcal{U} \times J$ of $(f, (\lambda_0, x_0))$ in $\mathcal{F} \times (\Lambda \times S)$. We may identify $S'$ with $\mathbb{R}^{N-1}$. Define

$$R: \quad \mathcal{U} \times (J \cap (\Lambda \times S^K)) \quad \to \quad S'^K \times \mathcal{M}^K(\mathbb{R}, N-1), \tag{10.19}$$

$$R(g, \lambda, x) \quad := \quad (\Pi_0(g, \lambda, x) - x, \; D_x \Pi_0(g, \lambda, x)).$$

**10.6 Lemma :**

*Under the above assumptions, the derivative*

$$D_f R(f,\lambda_0,x_0): \qquad \mathcal{F} \quad \rightarrow \quad S'^K \times \mathcal{M}^K(\mathbb{R}, N-1) \tag{10.20}$$

*is surjective.*

**Proof :**

Except for the equivariance aspect, the proof is similar to [Fie2, §3]. In step 1 below we show that $D_f\Pi_0(f,\lambda_0,x_0): \mathcal{F} \rightarrow S'^K$ is surjective. In step 2 we show that $\{0\} \times \mathcal{M}^K(\mathbb{R}, N-1) \subseteq im\, D_f R(f,\lambda_0,x_0)$, completing the proof. For these steps we will freely use the following facts on the relation between $\Pi_0$ and the flow $\Psi_t$:

$$\Pi_0(g,\lambda,x) \quad = \quad h^{-1}\Psi_{\tau(g,\lambda,x)}(g,\lambda,x), \tag{10.21.a}$$

where $\tau(g,\lambda,x)$ is the "Poincaré time" associated to the "Poincaré map" $\hat{\Pi}_0$ from $S = S_0$ to $S_1$. Because $\tau(f,\lambda_0,x_0) = p_0/m$, $m := |H/K|$, we obtain the derivative

$$D_f\Pi_0(f,\lambda_0,x_0) \quad = \quad P_S\, h^{-1}\, D_f\, \Psi_{p_0/m}(f,\lambda_0,x_0), \tag{10.21.b}$$

where $P_S$ denotes orthogonal projection onto $S'$ along $\dot{x}_0(0)$. Similarly

$$D_f(D_x\Pi_0(f,\lambda_0,x_0))\cdot g \quad = \quad P_S\, h^{-1}\, D_f(D_x\Psi_{p_0/m}(f,\lambda_0,x)|_{S'})\cdot g, \tag{10.21.c}$$

if we choose $g$ such that

$$g(\lambda_0,x_0(t)) \quad = \quad 0, \qquad \text{for all } t. \tag{10.21.d}$$

**Step 1 :** $\quad D_f\Pi_0(f,\lambda_0,x_0): \quad \mathcal{F} \rightarrow S'^K \quad$ is surjective.

By (10.21.b) it is sufficient to show that

$$D_f\Psi_{p_0/m}(f,\lambda_0,x_0): \qquad \mathcal{F} \rightarrow X^K \qquad \text{is surjective.}$$

Let

$$y(t) \quad := \quad D_f\Psi_t(f,\lambda_0,x_0)\cdot g\,.$$

Then

$$\begin{aligned} \dot{y}(t) \quad &= \quad D_x f(\lambda_0,x_0(t))y(t) \;+\; g(\lambda_0,x_0(t)) \\ y(0) \quad &= \quad 0\,. \end{aligned} \tag{10.22}$$

Given any $\hat{y} \in X^K$, define $y_1(t) := (tm/p_0)\hat{y}$ and $\hat{g}(t) \in X^K$ by

$$\dot{y}_1(t) \quad = \quad D_x f(\lambda_0,x_0(t))y_1(t) \;+\; \hat{g}(t)$$

for $0 \le t \le p_0/m$. Next we construct a continuous map

$$\tilde{g}: \qquad \{(\lambda_0,x_0(t)) \mid 0 \le t \le p_0/m\} \quad \rightarrow \quad X^K$$

such that

$$t \quad \mapsto \quad \tilde{g}(\lambda_0, x_0(t))$$

has compact support in $(0, p_0/m)$, but is nevertheless close enough to $\hat{g}(t)$ in $L^1((0,p/m), X^K)$ to guarantee

$$|y(p_0/m) - \hat{y}| \quad = \quad |y(p_0/m) - y_1(p_0/m)| \quad < \quad \varepsilon, \tag{10.23}$$

where $y$ solves equation (10.22) for $\tilde{g}$ instead of $g$. This is possible by the variation of constants formula for the linear inhomogeneous equation (10.22). We may extend $\tilde{g}$ to $\Lambda \times X$ such that $\tilde{g} \equiv 0$ outside a very small neighborhood of its already defined support in $\{(\lambda_0, x_0(t)) \mid 0 < t < p_0/m\}$. We may even perturb $\tilde{g}$ further to become smooth, keeping (10.23) valid.

It is the point of our whole detour via $\Pi_0$ that this nonlinearity $\tilde{g}$ can now be made $G$-equivariant on $\Lambda \times X$. Just define

$$g(\lambda, x) \quad := \quad \frac{1}{|K|} \sum_{\gamma \in G} \gamma^{-1} \tilde{g}(\lambda, \gamma x) \ ,$$

to obtain an equivariant $g$. By $G_{x_0(t)} = K$ and the above construction of the support, $g$ coincides with $\tilde{g}$ on $\{(\lambda_0, x_0(t)) \mid 0 \le t \le p_0/m\}$. In other words, $y(t)$ solves (10.22) for $g$ and $\tilde{g}$ alike.

By (10.23), $im\ D_f \Psi_{p_0/m}(f, \lambda_0, x_0)$ is therefore dense in $X^K$. Because $dim\ X^K$ is finite, this implies that $D_f \Psi_{p/m}(f, \lambda_0, x_0)$ is indeed surjective.

**Step 2 :** $\quad \{0\} \times \mathcal{M}^K(\mathbb{R}, N-1) \quad \subseteq \quad im\ D_f R(f, \lambda_0, x_0)$.

Restricting to $g$ with $g(\lambda_0, x_0(t)) = 0$ for all $t$, fact (10.21.c) makes it sufficient to show surjectivity of

$$g \quad \mapsto \quad h^{-1} D_f(D_x \Psi_{p_0/m}(f, \lambda_0, x_0)|_{S'}) \cdot g \ \in \ \mathcal{M}^K(\mathbb{R}, N \times (N-1)) \tag{10.24}$$

for this restricted class of $g$. Here $\mathcal{M}^K(\mathbb{R}, N \times (N-1))$ denotes the real $K$-equivariant $N \times (N-1)$ matrices; we identify $\mathbb{R}^N = X$, $\mathbb{R}^{N-1} = S'$. Let

$$\begin{aligned} S'_t \quad &:= \quad D_x \Psi_t(f, \lambda_0, x_0) S' \ , \\ S_t \quad &:= \quad \{x_0(t)\} + S'_t \\ Z(t) \quad &:= \quad D_f(D_x \Psi_t(f, \lambda_0, x_0)|_{S'}) \cdot g \end{aligned}$$

A direct calculation shows that $g(\lambda_0, x_0(t)) \equiv 0$ implies

$$\begin{aligned} \dot{Z}(t) \quad &= \quad D_x f(\lambda_0, x_0(t)) Z(t) \quad + \quad D_x g(\lambda_0, x_0(t)) \cdot (D_x \Psi_t(f, \lambda_0, x_0)|_{S'}) \ ; \\ Z(0) \quad &= \quad 0 \ . \end{aligned}$$

We discuss solutions of this linear inhomogeneous equation much in the spirit of step 1. Given $\hat{Z} \in \mathcal{M}^K(\mathbb{R}, N \times (N-1))$ let $Z_1(t) := (tm/p_0)\hat{Z}$, and define $\hat{g}(t) \in \mathcal{M}^K(\mathbb{R}, N \times (N-1))$ by

$$\dot{Z}_1(t) \quad = \quad D_x f(\lambda_0, x_0(t)) Z_1(t) \quad + \quad \hat{g}(t)$$

for $0 \le t \le p_0/m$. For some given small $\varepsilon > 0$ we now choose $g \in C^1$ in a neighborhood of $\{(\lambda_0, x_0(t)) \mid 0 \le t \le p_0/m\}$ such that $g(\lambda_0, x_0(t)) \equiv 0$ and $t \mapsto D_x g(\lambda_0, x_0(t))|_{S'_t}$ has compact support in $(0, p_0/m)$. Moreover $D_x g(\lambda_0, x_0(t)) \cdot (D_x \Psi_t(f, \lambda_0, x_0)|_{S'})$ is required to be close enough to $\hat{g}(t)$ in $L^1((0, p_0/m), \mathcal{M}^K(\mathbb{R}, N \times (N-1)))$ to guarantee

$$|Z(p/m) - Z_1(p/m)| \quad = \quad |Z(p/m) - \hat{Z}| \quad < \quad \varepsilon \, . \tag{10.25}$$

As in step 1, $g$ may be taken to be smooth. More to the point, $g$ extends $G$-equivariantly as before. Note that $D_x g(\lambda_0, x_0(t))$ is defined consistently with the requirement $g(\lambda_0, x_0(t)) \equiv 0$ because $\dot{x}_0(t) \notin S'_t$ and because $K$ leaves both $\langle \dot{x}_0(t) \rangle$ and $S'_t$ fixed.

As in step 1, (10.25) implies surjectivity of (10.24) and the proof is complete. □

## 10.6 Reduction to minimal period

Let $G$ be finite. To construct the sets $\mathcal{K}, \mathcal{J}$ for an application of theorem 10.2 we first introduce plenty of notation. Let $\mathcal{B} \subseteq \Lambda \times X$ be a closed ball around (0,0) of arbitrarily large, but fixed, radius. Given $T_0 > 0$ and $f_0 \in \mathcal{G}_0(\mathcal{B})$, let $\mathcal{N}_0$ be the neighborhood of stationary solutions and $\mathcal{U}_0$ the neighborhood of $f_0$ constructed in lemma 10.5. For $f \in \mathcal{U}_0$, $0 < \mathbf{h_0} \leq |G| + 1$ and $k \geq 3$ we define the sets

$\mathcal{Q}(f, T, \mathbf{h_0})$ : all $(\lambda, x) \in \mathcal{B} \setminus \mathcal{N}_0$ such that
$\Psi_t(f, \lambda, x) \in \mathcal{B}$ for all $t$, $\Psi_t(f, \lambda, x)$ is periodic with minimal period $p$, symmetry $(H, K, \Theta)$; and (10.26.$a$)
$p \leq T$, or $T \leq p \leq 1.5T$ and $|H| \geq \mathbf{h_0}$ . (10.26.$b$)

$\mathcal{P}(f, T, \mathbf{h_0})$ : all $(\lambda, x) \in \mathcal{Q}(f, T, \mathbf{h_0})$ such that

$$T \leq p \leq 1.5T \quad \text{and} \quad |H| = \mathbf{h_0} \,. \tag{10.27}$$

$\mathcal{O}_k(f)$ : all $(\lambda, x) \in \mathcal{B}$, satisfying (10.26.$a$), such that
$(\lambda, x)$ is of type $\mathcal{O}_k$, i.e. no primitive $e$-th root of unity, $1 \leq e \leq k$, is an eigenvalue of $\Pi_0$, or (10.28.$a$)
$(\lambda, x)$ is one of the non-hyperbolic bifurcation orbits given in lemma 5.9 (see also table 5.1, (c) and definition 5.10). (10.28.$b$)

The set $\mathcal{O}_k(f)$ is pronounced "okay $f$". Finally define

$\mathcal{G}_k(T, \mathbf{h_0})$ : all $f \in \mathcal{U}_0$ such that

$$\mathcal{Q}(f, T, \mathbf{h_0}) \subseteq \mathcal{O}_k(f) \,. \tag{10.29}$$

$\mathcal{G}(T_0)$ : all $f \in \mathcal{F}$ such that
(10.28.$b$) holds for all non-hyperbolic periodic solutions $(\lambda, x(t))$ with minimal period $\leq T_0$ and with $(\lambda, x(t)) \in \mathcal{B}$, for all $t$.

Let us compare the sets $\mathcal{P}$ and $\mathcal{Q}$. Suppose that $\Psi_{\tilde{p}}(f, \lambda, x) = x$ for some $(\lambda, x) \in \mathcal{Q}(f, T, \mathbf{h_0})$ and some $\tilde{p} \leq 1.5T$. Then $\tilde{p}$ need not be the minimal period of $(\lambda, x)$, and likewise we have no control on $|H|$ for the symmetry $(H, K, \Theta)$. Anyhow, $\mathcal{Q}(f, T, \mathbf{h_0})$ is compact. Indeed, passing to a limit, minimal periods never increase and $|H|$ never decreases, cf. §4. Considering $\mathcal{P}$ next, suppose that $\Psi_{\tilde{p}}(f, \lambda, x) = x$ for some $(\lambda, x) \in \mathcal{P}(f, T, \mathbf{h_0})$ and $\tilde{p} \leq 1.5T$. Then $\tilde{p}$ is in fact the minimal period, and $|H| = \mathbf{h_0}$. Unfortunately, $\mathcal{P}(f, T, \mathbf{h_0})$ need not be compact, e.g. due to doublings where $p$ jumps, and due to pitchforks which break $H$. Hence theorem 10.2 does not apply to $\mathcal{P}$ directly.

We will use an induction, instead, to show that $\mathcal{G}_k(T, \mathbf{h_0})$ is open dense in $\mathcal{U}_0$. There is a uniform lower bound, say $2\tau$, on minimal periods in $\mathcal{B}$ ([Las&Y]). Hence $f \in \mathcal{U}_0$ implies $\mathcal{Q}(f, \tau, 1) = \emptyset$, and therefore

$$\mathcal{G}_k(\tau, 1) = \mathcal{U}_0 \,. \tag{10.30.$a$}$$

From the definitions, it is obvious that $\mathcal{Q}(f, T, |G| + 1) = \mathcal{Q}(f, \frac{2}{3}T, 1)$, and therefore

$$\mathcal{G}_k(T, |G| + 1) = \mathcal{G}_k(\frac{2}{3}T, 1) \,. \tag{10.30.$b$}$$

The proof of lemma 5.9 reduces to the following crucial induction step.

$$\begin{array}{l}\text{If } \mathcal{G}_k(T,\mathbf{h_o}) \text{ is open dense in } \mathcal{U}_0,\ k > T/\tau, \\ \text{then } \mathcal{G}_k(T,\mathbf{h_o}-1) \text{ is also open dense in } \mathcal{U}_0 .\end{array} \tag{10.30.c}$$

Suppose this can be proved. Then, in particular, for $k > 2T_0/3\tau$

$$\mathcal{G}_k(\tfrac{2}{3}T_0, 1) \text{ is open dense in } \mathcal{U}_0 .$$

Because all periodic solutions in $\mathcal{N}_0$ with minimal period $\leq T_0$ are hyperbolic anyway, by lemma 10.5, this implies that

$$\mathcal{G}(T_0) \cap \mathcal{U}_0 \quad = \quad \bigcap_{k>2T_0/3\tau} \mathcal{G}_k(\frac{2}{3}T_0, 1)$$

is generic in $\mathcal{U}_0$. Because $\mathcal{U}_0$ was a neighborhood of $f_0$ in the open dense set $\mathcal{G}_0(\mathcal{B})$, we conclude that $\mathcal{G}(T_0)$ itself is generic. Because $T_0$ was arbitrarily large,

$$\bigcap_{T_0=1}^{\infty} \mathcal{G}(T_0) \qquad \text{is generic}.$$

By the "localization" described in §10.4, this proves lemma 5.9. Summarizing: the proof of lemma 5.9 hinges on the induction step (10.30.c).

Before we prove (10.30.c), we collect some continuity properties.

**10.7 Lemma :**

*For any $f_0 \in \mathcal{F}$ and for any open neighborhood $\mathcal{N}$ of $\mathcal{Q}(f_0, T, \mathbf{h_o})$ there exists a neighborhood $\mathcal{U}$ of $f_0$ such that*

$$\mathcal{Q}(f, T, \mathbf{h_o}) \subseteq \mathcal{N}, \qquad \text{for all } \ f \in \mathcal{U} . \tag{10.31}$$

**Proof :**

On the contrary, assume there exists an open neighborhood $\mathcal{N}$ of $\mathcal{Q}(f_0, T, \mathbf{h_o})$ and sequences

$$\begin{array}{ccc} & f_n \to f_0 & \\ (\lambda_n, x_n) & \in & \mathcal{Q}(f_n, T, \mathbf{h_o}) \setminus \mathcal{N}, \\ p_n \leq 1.5T, & & (H_n, K_n, \Theta_n) . \end{array}$$

By compactness we may assume convergence

$$\begin{array}{rcl} (\lambda_n, x_n) & \to & (\lambda_0, x_0) \ \in \ \mathcal{B} \setminus (\mathcal{N}_0 \cup \mathcal{N}), \\ p_n & \to & p_\infty \\ (H_n, K_n, \Theta_n) & = & (H_\infty, K_\infty, \Theta_\infty) . \end{array}$$

Here we use that $H_n \leq G$ and $G$ is finite. For $p, (H, K, \Theta)$ associated to $(\lambda_0, x_0)$, this implies

$$p \ = \ p_\infty \ \leq \ 1.5T, \qquad |H| \geq |H_\infty| , \quad \text{or}$$

$$p \ \leq \ 0.5 p_\infty \ \leq \ 0.75T ,$$

i.e. $(\lambda_0, x_0) \in \mathcal{Q}(f_0, T, \mathbf{h_o}) \subseteq \mathcal{N}$. This contradiction proves the lemma. □

**10.8 Lemma :**

*Assume that $f_0 \in \mathcal{G}_k(T,\mathbf{h_o})$ for some $k > T/\tau$ and that $\mathcal{G}_k(T,\mathbf{h_o})$ is open. Then there exist open neighborhoods $\mathcal{U}$ of $f_0$ in $\mathcal{G}_k(T,\mathbf{h_o})$ and $\mathcal{N}$ of $\mathcal{Q}(f_0,T,\mathbf{h_o})$ such that for any $f \in \mathcal{U}$ the following holds*

$$\begin{aligned} \mathcal{Q}(f,T,\mathbf{h_o}) &\subseteq \mathcal{N} & (10.32.a)\\ \mathcal{Q}(f,T,\mathbf{h_o}-1)\cap\mathcal{N} &\subseteq \mathcal{O}_k(f) & (10.32.b)\\ \mathcal{Q}(f,T,\mathbf{h_o}-1)\setminus\mathcal{N} &= \mathcal{P}(f,T,\mathbf{h_o}-1)\setminus\mathcal{N}. & (10.32.c)\end{aligned}$$

**Proof :**

By lemma 10.7, (10.32.$a$) is clear. By definition, (10.32.$a$) implies (10.32.$c$). It remains to prove (10.32.$b$). By compactness of $\mathcal{Q}(f_0,T,\mathbf{h_o})$ it is sufficient to prove the following local version of (10.32.$b$).

For any $(\lambda_0,x_0) \in \mathcal{Q}(f_0,T,\mathbf{h_o})$ there exist neighborhoods $\mathcal{U}'$ of $f_0$ in $\mathcal{G}_k(T,\mathbf{h_o})$ and $\mathcal{N}'$ of $(\lambda_0,x_0)$ such that for any $f \in \mathcal{U}'$

$$\mathcal{Q}(f,T,\mathbf{h_o}-1)\cap\mathcal{N}' \subseteq \mathcal{O}_k(f). \qquad (10.32.b)_{loc}$$

But indeed the local bifurcation picture near $(\lambda_0,x_0) \in \mathcal{Q}(f_0,T,\mathbf{h_o}) \subseteq \mathcal{O}_k(f)$ persists under small perturbations of $f_0$ by the implicit function theorem.

However, note that $(\lambda_0,x_0)$ may have Floquet multipliers which are primitive $e$-th roots of unity for some $e > k$. This does not introduce additional bifurcations within $\mathcal{Q}(f,T,\mathbf{h_o})$, or else the bifurcation orbit $(\lambda_0,x_0)$ would have period

$$p \leq 1.5T/k < 2\tau,$$

by virtual period corollary 4.6 and lemma 4.8. But this is impossible because $2\tau$ is a lower bound on the minimal periods.

Because secondary branches at the generic bifurcation points $(\lambda_0,x_0)$ out of table 5.1 are always hyperbolic, locally, we can thus accomplish $(10.32.b)_{loc}$ by the implicit function theorem argument mentioned above. This completes our proof. □

## 10.7 Proof of lemma 5.9

In the last section we have already reduced the proof of lemma 5.9 to claim (10.30.$c$). Because $\mathcal{G}_k(T,\mathbf{h_o}-1) \subseteq \mathcal{G}_k(T,\mathbf{h_o})$ and because $\mathcal{G}_k(T,\mathbf{h_o})$ is assumed to be open dense, it is sufficient to show the following local version of (10.30.$c$).

For any $f_0 \in \mathcal{G}_k(T,\mathbf{h_o})$ there exists a neighborhood $\mathcal{U}$ of $f_0$ in $\mathcal{G}_k(T,\mathbf{h_o})$ such that $\mathcal{U}\cap\mathcal{G}_k(T,\mathbf{h_o}-1)$ is open dense in $\mathcal{U}$. (10.33)

We construct $\mathcal{U}$ as follows. By lemma 10.8, there exist open neighborhoods $\mathcal{U}'$ of $f_0$ in $\mathcal{G}_k(T,\mathbf{h_o})$ and $\mathcal{N}'$ of $\mathcal{Q}(f_0,T,\mathbf{h_o})$ such that

$$\begin{aligned} \mathcal{U}'\cap\mathcal{G}_k(T,\mathbf{h_o}-1) &= \left\{f\in\mathcal{U}' \mid \mathcal{Q}(f,T,\mathbf{h_o}-1) \subseteq \mathcal{O}_k(f)\right\}\\ &= \left\{f\in\mathcal{U}' \mid (\mathcal{Q}(f,T,\mathbf{h_o}-1)\setminus\mathcal{N}') \subseteq \mathcal{O}_k(f)\right\} \qquad (10.34)\\ &= \left\{f\in\mathcal{U}' \mid (\mathcal{P}(f,T,\mathbf{h_o}-1)\setminus\mathcal{N}') \subseteq \mathcal{O}_k(f)\right\}.\end{aligned}$$

From §4 we conclude: for each $(\lambda_0, x_0) \in \mathcal{P}(f_0, T, \mathbf{h_0}-1) \setminus \mathcal{N}'$ with associated $p, (H, K, \Theta)$ there exist a neighborhood $\mathcal{U}(\lambda_0, x_0)$ of $f_0$ in $\mathcal{U}'$ and a neighborhood $\mathcal{J}(\lambda_0, x_0)$ of $(\lambda_0, x_0)$ in $\mathcal{B}$ such that the following holds.

$\mathcal{J}(\lambda_0, x_0)$ admits a fixed Poincaré section $S = \{x_0\} + \langle \dot{x}_0 \rangle^{\perp}$ for all $f \in \mathcal{U}(\lambda_0, x_0)$; (10.35.a)

for each $f \in \mathcal{U}(\lambda_0, x_0)$, and for each $(\lambda, x) \in ((\mathcal{P}(f, T, \mathbf{h_0}-1) \setminus \mathcal{N}') \cap \mathcal{J}(\lambda_0, x_0))$, the symmetry of $(\lambda, x)$ is again $(H, K, \Theta), |H| = \mathbf{h_0} - 1$, and the minimal period $p$ lies in $[T, 1.5T]$. (10.35.b)

Note that $\Pi_0 = \Pi_0(f, \lambda, x)$ is now well-defined on $\mathcal{U}(\lambda_0, x_0) \times \mathcal{N}(\lambda_0, x_0)$. Choose a closed neighborhood $\mathcal{K}(\lambda_0, x_0)$ of $(\lambda_0, x_0)$ in $\mathcal{J}(\lambda_0, x_0)$. Fortunately, $\mathcal{P}(f_0, T, \mathbf{h_0}-1) \setminus \mathcal{N}'$ is compact. We select a finite cover $\hat{\mathcal{K}} := \bigcup_j \mathit{int}\ \mathcal{K}(\lambda_j, x_j)$. Applying lemma 10.7 a last time to the neighborhood

$$\mathcal{N}^* \quad := \quad \mathcal{N}' \cup \bigcup_j \mathit{int}\ \mathcal{K}(\lambda_j, x_j)$$

of $\mathcal{Q}(f, T, \mathbf{h_0}-1)$, we obtain a neighborhood

$$\mathcal{U} \quad := \quad \bigcap_j \mathcal{U}(\lambda_j, x_j) \quad \subseteq \quad \mathcal{U}' \quad \subseteq \quad \mathcal{G}_k(T, \mathbf{h_0})$$

of $f_0$ in $\mathcal{G}_k(T, \mathbf{h_0})$. We claim that (10.33) holds for this choice of $\mathcal{U}$.

Given our interwoven construction of sets, we can now harvest the "fruit of the loom". As in (10.19) define

$$R \quad := \quad (\Pi_0 - x,\ D_x \Pi_0)$$

for $J := \mathcal{J}(\lambda_j, x_j)$, $\mathcal{U}$ as above, $\mathcal{Y} := S'^K \times \mathcal{M}^K(\mathbf{R}, N-1)$. We apply theorem 10.2 successively to various varieties $V = \{0\} \times V'$ given below; this part will be similar to the proof of theorem 5.2. Of course, $R$ is Fredholm with index $i = 1 - \mathit{dim}\ \mathcal{M}^K(\mathbf{R}, N-1)$, proper on $\mathcal{K} := \mathcal{K}(\lambda_j, x_j) \cap (\Lambda \times S^K)$, and $C^3$. Hence regularity assumption (10.6) holds. To prove transversality condition (10.5), just apply perturbation lemma 10.6. Note that the assumptions of lemma 10.6 on $\Pi_0$ and in particular on the precise symmetry of any orbit $(f, \lambda, x) \in \mathcal{U} \times \mathcal{K}$ are satisfied by (10.35.a, b) above. This was the reason for the whole inductive construction.

To finish the proof, we look up codimensions in proposition 10.3. With $V' = \mathcal{M}^K(\mathbf{R}, N-1)$, $\Pi_0$-fixed points in $\Lambda \times S^K$ are seen to be an embedded one-dimensional submanifold of $\Lambda \times S^K$, generically. Putting $V' = \mathcal{M}_{2,1}$ and $V' = \mathcal{M}'_{2,1}(k)$ next, the associated configurations of eigenvalues of $D_x \Pi_0$ are avoided, generically, because $\mathit{codim}\ V' = 2$. Putting $V' = \mathcal{M}_{1,1}$ we get simple eigenvalues $\pm 1$ of $D_x \Pi_0$ together with transversality conditions. In particular, the representation of $K$ on the eigenspace has to be real one-dimensional. This yields turn, flop, flip, or flip-flop; except for the curvature conditions.

Obviously the curvature conditions are open in the $C^3$-topology. To prove density, we construct a local perturbation of $f$. For the turn, we may perturb in $X^K$ as in the case of no symmetry, cf. [Fie2]. In all other cases, equivariant Ljapunov-Schmidt reduction [Van1] yields a one-dimensional kernel with an equation of the form

$$0 \quad = \quad x \cdot a(\lambda, x), \qquad\qquad x \in \mathbf{R}.$$

In the flop cases, $a(\lambda, x)$ is even in $x$ by equivariance. For the flip, $a(\lambda, x)$ can also be assumed to be even, after a suitable reparametrization of $x$; cf. (3.10.c). For an open dense set of $f$,

$$a(\lambda, x) \quad = \quad a_0(\lambda) \; + \; a_2(\lambda)x^2 \; + \; h.o.t. \; ,$$

$a_0(\lambda_0) = 0$, $a_2(\lambda_0) \neq 0$, and the curvature condition is satisfied each time. We skip further details.

Thus an application of theorem 10.2 proves that indeed

$$\mathcal{Q}(f, T, \mathbf{h_0} - 1) \setminus \mathcal{N}' \quad = \quad \mathcal{P}(f, T, \mathbf{h_0} - 1) \setminus \mathcal{N}' \quad \subseteq \quad \mathcal{O}_k(f)$$

for an open dense subset of $f \in \mathcal{U}$. On the other hand $\mathcal{Q}(f, T, \mathbf{h_0} - 1) \cap \mathcal{N}' \subseteq \mathcal{O}_k(f)$, as we recall from $(10.32.b)_{loc}$. This completes the induction step (10.33), (10.30.c) and the proof of lemma 5.9. □

The proofs of theorem 5.11 (a,b) will be even easier.

## 10.8 Proof of theorem 5.11 (a)

We consider periodic solutions with symmetry $(H, K, \Theta)$, $K = H = G = \mathbb{Z}_\infty$. These solutions lie in the flow-invariant subspace $X^G$, where the action of $G$ is trivial. Restricting to generic bifurcations within $X^G$ we may therefore invoke the standard generic results for the non-equivariant case [All&M-P&Y2, Fie2, Med, Sot]. They tell us that, generically, only turns and flip doublings occur within $X^G$. Alternatively we could revive our proof of lemma 5.9 specializing the induction over minimal period and symmetry to periodic solutions in $X^G$.

Still something remains to be checked. In theorem 5.11 we claim that table 5.2 lists **all** generic secondary bifurcations. We still have to prove that there are no symmetry breaking bifurcations from $(\lambda_0, x_0) \in \Lambda \times X^G$ to periodic solutions in $X \setminus X^G$, generically.

To prove this we employ the period induction from the proof of lemma 5.9, cf. §§10.6, 10.7. We just modify the sets $\mathcal{Q}(f, T)$, $\mathcal{P}(f, T)$ to consist of periodic solutions in $X^G$, only, omitting $\mathbf{h_0} = |H|$ altogether. Similarly, $\mathcal{O}_k(f)$ should refer to periodic solutions in table 5.2(a) now. For generic $f \in \mathcal{F}$ we conclude that the only relevant Floquet multipliers are simple eigenvalues $+1$ and $-1$, generically. If the corresponding eigenvector of the linearized, $G$-equivariant Poincaré map

$$D_x\Pi(\lambda_0, x_0) : \qquad S' \to S'$$

lies in $S' \cap X^G$, then we obtain a turn respectively a flip doubling within $X^G$. But if this eigenvector was in $S' \setminus X^G$ then, acting on it with $G = \mathbb{Z}_\infty$, the corresponding Floquet multiplier could not be simple. Therefore we see only turns and flip doublings within $X^G$, generically. This completes the proof of theorem 5.11 (a). □

## 10.9 Proof of theorem 5.11 (b)

We consider rotating and frozen waves $(\lambda_0, x_0(t))$. By §5.3 they satisfy

$$-\alpha \mathcal{R} x_0 \; + \; f(\lambda, x_0) \quad = \quad 0 \tag{5.3}$$

with $\mathcal{R}x_0 \neq 0$. We will first analyse frozen waves ($\alpha = 0$), showing that any frozen wave is a freezing, generically. Because

$$\alpha = (\Theta p)^{-1} \tag{5.4}$$

for rotating waves ($\alpha \neq 0$) it then remains to discuss generic rotating waves, with some lower bound on $|\alpha|$ or, equivalently, some upper bound on the minimal period $p$. Again we will rely on a (modified) period induction similar to §§10.6, 10.7.

Let us prove that any frozen wave $(\lambda_0, x_0)$ is a freezing, generically. From §5.3 we recall the Ljapunov-Schmidt reduction

$$-\alpha Q\mathcal{R}x + Qf(\lambda, x) = 0 \quad \Longleftrightarrow \quad \alpha = \alpha(f, \lambda, x) \tag*{(5.3)$_Q$}$$

$$\mathcal{T}(f, \lambda, x) := P\Big(-\alpha(f, \lambda, x)\mathcal{R}x + f(\lambda, x)\Big) = 0 \tag*{(5.3)$_P$}$$

near a frozen wave. We apply transversality theorem 10.2, again, with

$$R(f, w) := \Big(\alpha(f, \lambda, x), \mathcal{T}(f, \lambda, x), \; D_x\mathcal{T}(f, \lambda, x)\Big), \qquad w := (\lambda, x). \tag{10.36}$$

Note that $\alpha$, $\mathcal{T}$, and hence $R$, are defined only locally. By the usual compactness argument, this will not be a restriction. We consider $R$ as a map

$$R: \qquad (\mathcal{F} \times \Lambda \times S^K)_{loc} \longrightarrow \boldsymbol{R} \times S'^K \times \mathcal{M}^K(\boldsymbol{R}, N-1),$$

where $K := G_{x_0}$ denotes the isotropy of $x_0$, cf. §10.5, (10.19).

Preparing another application of transversality theorem 10.2, we claim that perturbation lemma 10.6 still holds, i.e.

$$D_f R(f, \lambda_0, x_0): \qquad \mathcal{F} \rightarrow \boldsymbol{R} \times S'^K \times \mathcal{M}^K(\boldsymbol{R}, N-1) \tag{10.20}$$

is still surjective. Indeed, for $\alpha = 0$,

$$D_f R(f, \lambda_0, x_0) \cdot g = \Big(D_f\alpha(f, \lambda_0, x_0) \cdot g, \; Pg(\lambda_0, x_0), \; PD_x g(\lambda_0, x_0)\Big)$$

can be prescribed arbitrarily, choosing an appropriate $G$-equivariant $g \in \mathcal{F}$ as follows. First we define a $K$-equivariant $\tilde{g}$ on $(\Lambda \times S)_{loc}$ using a $K$-invariant $C^\infty$ cut-off in $S$ near $(\lambda_0, x_0)$. In particular, we can prescribe $D_f\alpha(f, \lambda_0, x_0) \cdot \tilde{g}$ because

$$(D_f\alpha(f, \lambda_0, x_0) \cdot \tilde{g}) \, Q\mathcal{R}x_0 = Q\tilde{g}(\lambda_0, x_0),$$

and because $Q\mathcal{R}x_0 = \mathcal{R}x_0 \neq 0$ by construction. We then extend $\tilde{g}$ globally, defining

$$g(\lambda, x) := \rho(h)\, \tilde{g}(\lambda, \rho(h)^{-1}x),$$

if the group orbit $Gx$ hits $S$; $g(\lambda, x) := 0$ otherwise. Note that $g$ is well-defined by $K$-equivariance of $\tilde{g}$ on $S$. This proves lemma 10.6 for $R$ as above.

We apply transversality theorem 10.2 to

$$V := \boldsymbol{R} \times \{0\} \times \mathcal{M}^K(\boldsymbol{R}, N-1),$$

first. Now, all assumptions hold. We conclude that for generic $f$ the solutions $(\lambda, x) \in \Lambda \times S^K$ of

$$\mathcal{T}(f, \lambda, x) = 0$$

form a $C^4$ branch $(\lambda, x_\lambda)$ through $(\lambda_0, x_0)$, locally. Putting

$$V := \{0\} \times \{0\} \times \mathcal{M}^K(\mathbb{R}, N-1)$$

next, we see that

$$\lambda \mapsto \alpha(f, \lambda, x_\lambda)$$

has only simple zeros, generically. This takes care of transversality condition (5.7.$b$) for a freezing. To show genericity of the hyperbolicity condition (5.7.$a$)$'$ we put

$$V := \{0\} \times \{0\} \times \mathcal{M}_{1,0}$$

Then

$$\mathcal{R}(f, \cdot, \cdot) \cap V = \emptyset,$$

for generic $f$, because assumption (10.8.$b$) holds:

$$\begin{aligned} i + \dim V &= (-1 + 1 - \dim \mathcal{M}^K(\mathbb{R}, N-1)) + \dim \mathcal{M}_{1,0} = \\ &= -\operatorname{codim} \mathcal{M}_{1,0} = -1 < 0, \end{aligned}$$

using $\operatorname{codim} \mathcal{M}_{1,0} = 1$ from proposition 10.3. Therefore $D_x \mathcal{T}(f, \lambda_0, x_0)$ is indeed hyperbolic, as in (5.7.$a$)$'$, at a generic frozen wave. This proves that any frozen wave is a freezing, generically.

Let us now turn to rotating waves ($\alpha \neq 0$). Before we can apply transversality theorem 10.2, we modify the setting of our period induction §§10.6, 10.7, slightly. Working within a large closed ball $\mathcal{B}$, as in §10.6, we admit as elements of the sets $\mathcal{Q}$ and $\mathcal{P}$ only periodic solutions $(\lambda, x) \in \mathcal{B} \setminus \mathcal{N}_0$ with symmetry $(H, K, \Theta)$, $H = \mathbb{Z}_\infty$, $\Theta \neq 0$, i.e. we admit only rotating waves. Notationally, we replace

$$\left.\begin{array}{l} \mathcal{Q}(f, T, \mathbf{h_0}) \\ \mathcal{Q}(f, T, \mathbf{h_0} - 1) \\ \mathcal{P}(f, T, \mathbf{h_0}) \\ \mathcal{P}(f, T, \mathbf{h_0} - 1) \\ \mathcal{G}_k(T, \mathbf{h_0}) \\ \mathcal{G}_k(T, \mathbf{h_0} - 1) \end{array}\right\} \text{ by } \left\{\begin{array}{l} \mathcal{Q}(f, T) \\ \mathcal{Q}(f, \frac{2}{3}T) \\ \mathcal{P}(f, T) \\ \mathcal{P}(f, \frac{2}{3}T) \\ \mathcal{G}_k(T) \\ \mathcal{G}_k(\frac{2}{3}T) \end{array}\right.$$

Of course, $\mathcal{O}_k(f)$ then refers to bifurcations of rotating waves listed in table 5.2 (b). We also replace the induction step as follows.

If $\mathcal{G}_k(\frac{2}{3}T)$ is open dense in $\mathcal{U}_0$, $k > T/\tau$, then $\mathcal{G}_k(T)$ is also open dense in $\mathcal{U}_0$. (10.30.$c$)$'$

Then continuity lemmas 10.7, 10.8 hold verbatim, along with their proofs. The proof of lemma 5.9 applies up until the "fruit of the loom". In particular, we emphasize that all rotating waves in a small neighborhood $\mathcal{N}_0$ of the set of stationary solutions are hyperbolic, because such solutions arise either from centers or from freezings. Moreover, there is a uniform lower bound

$$|\alpha(f,\lambda,x)| \quad \geq \quad \alpha_0 \quad > \quad 0$$

for any rotating wave $(\lambda, x) \in \mathcal{B} \setminus \mathcal{N}_0$, as long as $f$ stays in some small neighborhood $\mathcal{U}$ of $f_0$.

We now modify $R$ slightly, defining

$$R(f,\lambda,x) \quad := \quad \frac{1}{\alpha(f,\lambda,x)}\left(\mathcal{T}(f,\lambda,x),\ D_x\mathcal{T}(f,\lambda,x)\right)$$

with the usual local domain of definition. As before, perturbation lemma 10.6 holds. We apply transversality theorem 10.2 to

$$V \quad := \quad \{0\} \times \mathcal{M}'_{2,0}\,,$$

first. Then

$$R(f,\cdot,\cdot) \quad \not\pitchfork \quad V\,,$$

for generic $f$, because assumption (10.8.b) holds:

$$i + \mathit{dim}\, V \quad = \quad 1 - \mathit{codim}\, \mathcal{M}'_{2,0} \quad = \quad -1 \quad < \quad 0\,,$$

using $\mathit{codim}\, \mathcal{M}'_{2,0} = 2$ from proposition 10.3. Eigenvalues of $D_x\mathcal{T}(f,\lambda_0,x_0)$ at a rotating wave are related to its Floquet multipliers by (5.5.b), putting $t = p$, $\alpha p = 1/\Theta$:

$$\begin{aligned} D_x\Pi(f,\lambda_0,x_0) \quad &= \quad P\,\rho(1/\Theta)\, \mathit{exp}\left(\left(-\mathcal{R} + \frac{1}{\alpha}D_x f(\lambda_0,x_0)\right)/\Theta\right)\Big|_{S'} \\ &= \quad \rho(1/\Theta)\, \mathit{exp}\left(\frac{1}{\Theta}\, D_x\mathcal{T}(f,\lambda_0,x_0)\right)\,. \end{aligned} \tag{10.37}$$

Indeed, this last equality holds because the orthogonal projection $P$ along $\mathcal{R}x_0$ commutes with $\rho(1/\Theta)$ and because

$$P(-\mathcal{R} + \frac{1}{\alpha}D_x f(\lambda_0,x_0)) \quad = \quad P(-\mathcal{R} + \frac{1}{\alpha}D_x f(\lambda_0,x_0))\,P$$

Thus $D_x\mathcal{T}(f,\lambda_0,x_0) \not\pitchfork \mathcal{M}'_{2,0}(k)$ implies that $D_x\Pi(f,\lambda_0,x_0))^{|\Theta|}$ does not possess an eigenvalue $\mu$ which is an $e$-th root of unity, $e \leq k$, except possibly for those coming from an eigenvalue 0 of $D_x\mathcal{T}(f,\lambda_0,x_0)$. Consequently, the same is true for eigenvalues $\mu_0$ of $D_x\Pi(f,\lambda_0,x_0)$ itself. Taking

$$V := \{0\} \times \mathcal{M}_{2,0}\,, \qquad V := \{0\} \times \mathcal{M}_{1,0}\,,$$

successively, any eigenvalue $\mu = 0$ has to be simple. Hence $K = \{0, 1/\Theta, \cdots, (\Theta-1)/\Theta\}$ acts by $\rho^{\pm}$ on the one-dimensional eigenspace. Taking

$$V \quad := \quad \{0\} \times \mathcal{M}^K(\mathbb{R}, N-1)\,,$$

the case of $\rho^+$ corresponds to a turn, because the zeros $(\lambda, x)$ of $\mathcal{T}(f,\cdot)$ yield one-dimensional submanifolds of $\Lambda \times S^K$, generically. The transversality condition (3.7.b) is satisfied generically, by the above choice of $V = \{0\} \times \mathcal{M}_{1,0}$. The curvature condition (3.7.c) is generic by an explicit perturbation, as discussed in the discrete case. The case of representation $\rho^-$ corresponds to a flop for $\mathcal{T}(f,\cdot,\cdot)$, generically, i.e. to a flop doubling of rotating waves, by analogous arguments.

Finally, we conclude from the linearized Poincaré map (10.37) that the above bifurcations, turn and flop doubling, constitute the only bifurcations of rotating waves (aside from freezing). In particular, symmetry breaking bifurcations to periodic solutions with symmetry $(\tilde{H},\tilde{K},\tilde{\Theta})$, $\tilde{H} < H = \mathbb{Z}_\infty$, are nongeneric. Indeed, in any of the above bifurcations, the local branches are determined uniquely by an application of the implicit function theorem. And all these branches consist of rotating waves, exclusively. This completes the proof of theorem 5.11 (b), the proof of theorem 5.11 as a whole, and the opus. □

# References

[Ab&Mars] R. Abraham & J. Marsden, *Foundations of Mechanics*, Benjamin/Cummings, Reading 1978.

[Ab&Ro] R. Abraham & J. Robbin, *Transversal Mappings and Flows*, Benjamin Inc., Amsterdam 1967.

[Abram&Ste] M. Abramowitz & I.A. Stegun (eds.), *Handbook of Mathematical Functions*, Dover Publications Inc., New York 1965.

[Agl&Kri] K.I. Agladze & V.I. Krinsky, *Multi-armed vortices in an active chemical medium*, Nature **296** (1982), 424-426.

[Ale1] J.C.Alexander, *Bifurcation of zeroes of parametrized functions*, J. Func. Anal. **29** (1978), 37-53.

[Ale2] ———, *Patterns at primary Hopf bifurcations of a plexus of identical oscillators*, SIAM J. Appl. Math. **46** (1986), 199-221.

[Ale3] ———, *Spontaneous oscillations in two 2-component cells coupled by diffusion*, J. Math. Biol. **23** (1986), 205-219.

[Ale4] ———, *A period-doubling bubble in the dynamics of two coupled oscillators*, in [Oth], 208-220.

[Ale&Au1] J.C. Alexander & J.F.G. Auchmuty, *Global branching of waves*, manus. math. **27** (1979), 159-166.

[Ale&Au2] ———, *Global bifurcations of phase-locked oscillators*, Arch. Rat. Mech. Anal. **93** (1986), 253-270.

[Ale&Fie] J.C. Alexander & B. Fiedler, *Global decoupling of coupled symmetric oscillators*, preprint 1987.

[Ale&Fitz] J.C. Alexander & P.M. Fitzpatrick, *The homotopy of certain spaces of nonlinear operators, and its relation to global bifurcation of the fixed points of parametrized condensing operators*, J. Func. Anal. **34** (1979), 87-106.

[Ale&Y1] J.C. Alexander & J.A. Yorke, *Global bifurcation of periodic orbits*, Am. J. Math. **100** (1978), 263-292.

[Ale&Y2] ———, *On the continuability of periodic orbits of parametrized three-dimensional differential equations*, J. Diff. Eq. **49** (1983), 171-184.

[All&M-P&Y1] K.T. Alligood & J. Mallet-Paret & J.A. Yorke, *Families of periodic orbits: local continuability does not imply global continuability*, J. Diff. Geom. **16** (1981), 483-492.

[All&M-P&Y2] ———, *An index for the global continuation of relatively isolated sets of periodic orbits*, in [Pal], 1-21.

[All&Y1] K.T. Alligood & J.A. Yorke, *Families of periodic orbits: virtual period and global continuability*, J. Diff. Eq. **55** (1984), 59-71.

[All&Y2] ———, *Hopf bifurcation: the appearance of virtual periods in cases of resonance*, J. Diff. Eq. **64** (1986), 375-394.

[And&Chai] A.A. Andronov & C.E. Chaikin, *Theory of Oscillations*, Princeton Univ. Press, Princeton 1949.

[And&Leo&Gor&Mai] A.A. Andronov & E.A. Leontovich & I.I. Gordon & A.G. Maier, *Theory of Bifurcations of Dynamic Systems on a Plane*, Israel Progr. Sci. Transl., Jerusalem 1973.

[Ang&Fie] S. Angenent & B. Fiedler, *The dynamics of rotating waves in scalar reaction diffusion equations*, Trans. AMS, to appear.

[Aris] R. Aris, *The Mathematical Theory of Diffusion and Reaction in Permeable Catalysts II*, Clarendon Press, Oxford 1975.

[Arn1] V.I. Arnold, *Lectures on bifurcations and versal systems*, Russ. Math. Surv. **27** (1972), 54-123.

[Arn2] ———, *Loss of stability of selfoscillations close to resonance and versal deformations of equivariant vector fields*, Func. Anal. Appl. **11** (1977), 85-92.

[Arn3] ———, *Geometrical Methods in the Theory of Ordinary Differential Equations*, Grundl. math. Wiss. 250, Springer-Verlag, New York 1983.

[Arn4] ———, *Catastrophe Theory*, Springer-Verlag, Berlin 1984.

[Arn&G-Z&Var] V.I. Arnold & S.M. Gusein-Zade & A.N. Varchenko, *Singularities of Differentiable Maps I*, Birkhäuser, Boston 1985.

[Aro&Doe&Oth] D.G. Aronson & E.J. Doedel & H.G. Othmer, *Bistable behavior in coupled oscillators*, in [Oth], 221-231.

[Ash&Oth] M. Ashkenazi & H.G. Othmer, *Spatial patterns in coupled bio-chemical oscillators*, J. Math. Biol. **5** (1978), 305-350.

[Au] J.F.G. Auchmuty, *Bifurcating waves*, in: Bifurcation Theory and Applications in Scientific Disciplines, O. Gurel & O.E. Rössler (eds.), Ann. NY Acad. Sc. **316**, New York 1979, 263-278.

[Bar] T. Bartsch, *Verzweigung in Vektorraumbündeln und äquivariante Verzweigung*, Dissertation, Univ. München 1986.

[Bog1] R.I. Bogdanov, *Bifurcations of a limit cycle of a family of vector fields in the plane*, russ., Trudy Sem. I.G. Petrovskogo **2** (1976), 23-26.

[Bog2] ———, *A versal deformation of a singular point of a vector field in the plane in the case of zero eigenvalues*, russ., Trudy Sem. I.G. Petrovskogo **2** (1976),37-65.

[Bre] G.E. Bredon, *Introduction to Compact Transformation Groups*, Academic Press, New York 1980.

[Brö&tD] T. Bröcker & T. tom Dieck, *Representations of Compact Lie Groups*, Grad. Texts in Math. 98, Springer-Verlag, New York 1985.

[Bru] P. Brunovský, *On one-parameter families of diffeomorphisms*, I and II, Comm. math. Univ. Carolinae **11** (1970), 559-582, and **12** (1971), 765-784.

[Cer] G. Cerami, *Symmetry breaking for a class of semilinear elliptic problems*, Nonl. Anal. TMA **10** (1986), 1-14.

[Cic] G. Cicogna, *Bifurcation from topology and symmetry arguments*, Bolletino U.M.I. Ser. 6, **3A** (1984), 131-138.

[Chos] P. Chossat, *Solutions avec symétry diédrale dans le problème de bifurcation invariants par symétry spherique*, C.R. Acad. Sci. I **297** (1983), 639-642.

[Chos&Dem&Io] P. Chossat & Y. Demay & G. Iooss, *Interaction de modes azimutaux dans le problème de Couette-Taylor*, Arch. Rat. Mech. Anal. **99** (1987), 213-248.

[Chos&Go&Key] P. Chossat & M. Golubitsky & B.L. Keyfitz, *Hopf-Hopf mode interactions with $O(2)$ symmetry, Dynamics and Stability of Systems*, to appear.

[Chow&Ha] S.-N. Chow & J.K. Hale, *Methods of Bifurcation Theory*, Grundl. math. Wiss. 251, Springer-Verlag, New York 1982.

[Chow&M-P] S.-N. Chow & J. Mallet-Paret, *The Fuller index and global Hopf bifurcation*, J. Diff. Eq. **29** (1978), 66-85.

[Chow&M-P&Y1] S.-N. Chow & J. Mallet-Paret & J.A. Yorke, *Global Hopf bifurcation from a multiple eigenvalue*, Nonlinear Analysis TMA **2** (1978), 753-763.

[Chow&M-P&Y2] ———, *A periodic orbit index which is a bifurcation invariant*, in [Pal], 109-131.

[Cra&Rab1] M.G. Crandall & P.H. Rabinowitz, *Bifurcation from simple eigenvalues*, J. Func. Anal. **8** (1971), 321-340.

[Cra&Rab2] ———, *The Hopf bifurcation theorem in infinite dimensions*, Arch. Rat. Mech. Anal. **67** (1977), 53-72.

[Cro&Field] M.F. Crowley & R.J. Field, *Electrically coupled Belousov-Zhabotinskii oscillators: experimental observations of chaos in a chemical system and identification of its source in the Field-Noyes equations*, in [Oth], 68-97.

[Cu&San] R. Cushman & J. Sanders, *Nilpotent normal forms and representation theory of $sl(2, R)$*, in [Go&Gu], 31-51.

[Da1] G. Dangelmayr, *Steady-state mode interactions in the presence of $O(2)$-symmetry*, Dynamics and Stability of Systems **1** (1986), 159-185.

[Da2] ———, *Wave patterns in coupled stationary bifurcations with $O(2)$-symmetry*, preprint 1986.

[Da&Arm] G. Dangelmayr & D. Armbruster, *Steady-state mode interactions in the presence of $O(2)$-symmetry and in no-flux boundary conditions*, in [Go&Gu], 53-68.

[Da&Kno] G. Dangelmayr & E. Knobloch, *The Takens-Bogdanov bifurcation with $O(2)$-symmetry*, preprint.

[Dei] K. Deimling, *Nonlinear Functional Analysis*, Springer-Verlag, Berlin 1985.

[Deu&Fie&Kun] P. Deuflhard & B. Fiedler & P. Kunkel, *Efficient numerical pathfollowing beyond critical points*, SIAM J. Numer. Anal. **24** (1987), 912-927.

[tD1] T. tom Dieck, *Transformation Groups and Representation Theory*, Lect. Notes Math. 766, Springer-Verlag, Berlin 1979.

[tD2] ———, *Transformation Groups*, de Gruyter, Berlin 1987.

[Di] J. Dieudonné, *Éléments d'Analyse 2*, Gauthier-Villars, Paris 1968.

[Doe&Ker1] E.J. Doedel & J.P. Kernevez, *Software for continuation problems in ordinary differential equations with applications*, California Institute of Technology 1985.

[Doe&Ker2] ———— , *A numerical analysis of wave phenomena in a reaction diffusion model*, in [Oth], 261-273.

[Do] A. Dold, *Lectures on Algebraic Topology*, Springer-Verlag, Heidelberg 1972.

[Dug] J. Dugundji, *Topology*, Allyn & Bacon Inc., Boston 1966.

[Ei&Schu] M. Eigen & P. Schuster, *The Hypercycle: A Principle of Natural Self-Organization*, Springer-Verlag, Berlin 1979.

[Ern&H-K] T. Erneux & M. Herschkowitz-Kaufman, *Rotating waves as asymptotic solutions of a model chemical reaction*, J. Chem. Phys. **66** (1977), 248-250.

[Fie1] B. Fiedler, *Global Hopf bifurcation in porous catalysts*, Equadiff 82, H.W. Knobloch & K. Schmitt (eds.), Lect. Notes Math. 1017, Springer-Verlag, Heidelberg 1983, 177-183.

[Fie2] ———— , *An index for global Hopf bifurcation in parabolic systems*, J. reine u. angew. Math. **359** (1985), 1-36.

[Fie3] ———— , *Global Hopf bifurcation of two-parameter flows*, Arch. Rat. Mech. Anal. **94** (1986), 59-81.

[Fie4] ———— , *Global Hopf bifurcation for Volterra integral equations*, SIAM J. Math. Analysis **17** (1986), 911-932.

[Fie5] ———— , *Global Hopf bifurcation in reaction diffusion systems with symmetry*, in: Dynamics of Infinite Dimensional Systems, S.-N. Chow & J.K. Hale (eds.), Springer-Verlag, Berlin 1987, 81-104.

[Fie&Kun1] B. Fiedler & P. Kunkel, *A quick multiparameter test for periodic solutions*, in [Kü&Sey&Tro].

[Fie&Kun2] ———— , *Multistability, scaling, and oscillations*, in: Modelling of Chemical Reaction Systems, Deuflhard et al. (eds.), Heidelberg 1986, to appear.

[Field] R.J. Field, *Experimental and mechanistic characterization of Bromate-ion-driven chemical oscillations and traveling waves in closed systems*, in [Field&Bur], 55-92.

[Field&Bur] R.J. Field & M. Burger (eds.), *Oscillations and Traveling Waves in Chemical Systems*, Wiley & Sons, New York 1985.

[Ful] F.B. Fuller, *An index of fixed point type for periodic orbits*, Am. J. Math. **89** (1967), 133-148.

[vG&M-P] S.A. van Gils & J. Mallet-Paret, *Hopf bifurcation and symmetry: travelling and standing waves on the circle*, Proc. Roy. Soc. Edinburgh **104A** (1986), 279-307.

[vG&Val] S.A. van Gils, T. Valkering, *Hopf bifurcation and symmetry: standing and travelling waves in a circular chain*, Japan. J. Appl. Math. **3** (1986), 207-222.

[Go&Gu] M. Golubitsky & J.M. Guckenheimer (eds.), *Multiparameter Bifurcation Theory*, Cont. Math. 56, AMS, Providence 1986.

[Go&Gui] M. Golubitsky & V. Guillemin, *Stable Mappings and Their Singularities*, Springer-Verlag, New York 1973.

[Go &Sch] M. Golubitsky & D.G. Schaeffer, *Singularities and Groups in Bifurcation Theory I*, Appl. Math. Sc. 51, Springer-Verlag, New York 1986.

[Go&Sch&St] M. Golubitsky & D.G. Schaeffer & I.N. Stewart, *Singularities and Groups in Bifurcation Theory II*, in preparation.

[Go&St1] M. Golubitsky & I.N. Stewart, *Hopf bifurcation in the presence of symmetry*, Arch. Rat. Mech. Anal. **87** (1985), 107-165.

[Go&St2] ———— , *Symmetry and stability in Taylor-Couette flow*, SIAM J. Math. Anal. **17** (1986), 249-288.

[Go&St3] ———— , *Hopf bifurcation with dihedral group symmetry: coupled nonlinear oscillators*, in [Go&Gu], 131-173.

[Go&Swi&Kno] M. Golubitsky & J.W. Swift & E. Knobloch, *Symmetries and pattern selection in Rayleigh-Bénard convection*, Physica **10D** (1984), 249-276.

[Gu] J. Guckenheimer, *A codimension two bifurcation with circular symmetry*, in [Go&Gu], 175-184.

[Gu&Ho] J. Guckenheimer & P. Holmes, *Nonlinear Oscillations, Dynamical Systems, and Bifurcations of Vector Fields*, Appl. Math. Sc. 42, Springer-Verlag, New York 1983.

[Harr&Y] J. Harrison & J.A. Yorke, *Flows on $S^3$ and $R^3$ without periodic orbits*, in [Pal], 401-407.

[Hart] P. Hartman, *Ordinary Differential Equations*, Birkhäuser, Boston 1982.

[Has&Kaz&Wan] B.D. Hassard & N.D. Kazarinoff & Y.-H. Wan, *Theory and Applications of the Hopf Bifurcation*, Cambridge Univ. Press, Cambridge 1980.

[Hasse] H. Hasse, *Number Theory*, Grundl. math. Wiss. 229, Springer-Verlag, Berlin 1980.

[Hau] H. Hauschild, *Zerspaltung äquivarianter Homotopiemengen*, Math. Ann. **230** (1977), 279-292.

[adH] U. an der Heiden, *Analysis of Neural Networks*, Lect. Notes Biomath. 35, Springer-Verlag 1980.

[Hen] D. Henry, *Geometric Theory of Semilinear Parabolic Equations*, Springer-Verlag, New York 1981.

[Hir1] M.W. Hirsch, *Differential Topology*, Springer-Verlag, New York 1976.

[Hir2] ———— , *Systems of differential equations that are competitive or cooperative II. Convergence almost everywhere*, SIAM J. Math. Anal. **16** (1985), 423-439.

[Hof&Sig] J. Hofbauer & K. Sigmund, *Evolutionstheorie und dynamische Systeme*, Verlag P. Parey, Berlin 1984.

[Hopf] E. Hopf, *Abzweigung einer periodischen Lösung von einer stationären Lösung eines Differential-Systems*, Ber. math.-phys. Klasse Sächs. Akad. Wiss. **94** (1942), 1-22.

[Hopp] F.R. Hoppensteadt, *Properties of solutions of ordinary differential equations with small parameters*, Comm. Pure Appl. Math. **24** (1971), 807-840.

[Hor] E.I. Horozov, *Versal deformations of equivariant vector fields for the cases of symmetries of order 2 and 3*, Trudy Sem. I.G. Petrovskogo **5** (1977), 163-192.

[How] L.N. Howard, *Nonlinear oscillations*, in: Nonlinear Oscillations in Biology, F.R. Hoppensteadt (ed.), AMS Lect. Math. Life Sc. **17** (1979), 1-69.

[Hub] A. Huber, *Über die Replikatorgleichung*, Diploma thesis, Heidelberg 1986.

[Ihr&Go] E. Ihrig & M. Golubitsky, *Pattern selection with $O(3)$ symmetry*, Physica **13D** (1984), 1-33.

[Io] G. Iooss, *Bifurcation of Maps and Applications*, Math. Stud. 36, North-Holland, Amsterdam 1979.

[Io&Jo] G. Iooss & D.D. Joseph, *Elementary Stability and Bifurcation Theory*, Springer-Verlag, New York 1980.

[Ize1] J. Ize, *Bifurcation Theory for Fredholm Operators*, AMS memoir 174, Providence 1976.

[Ize2] ———— , *Obstruction theory and multiparameter Hopf bifurcation*, Trans. AMS **289** (1985), 757-792.

[Ize&Mas&Pej&Vig] J. Ize & I. Massabó & J. Pejsachowicz & A. Vignoli, *Structure and dimension of global branches of solutions to multiparameter nonlinear equations*, Trans. AMS **291** (1985), 383-435.

[Ize&Mas&Vig] J. Ize & I. Massabó & A. Vignoli, *Global results on continuation and bifurcation for equivariant maps*, preprint 1985.

[Jep&Spe1] A.D. Jepson & A. Spence, *Singular points and their computation*, in [Kü&Mit&Web], 195-209.

[Jep&Spe2] ———— , *The numerical solution of nonlinear equations having several parameters I: scalar equations*, SIAM J. Numer. Anal. **22** (1985), 736-759.

[Kato] T. Kato, *Perturbation Theory for Linear Operators*, Springer-Verlag, New York 1980.

[Kie] H. Kielhöfer, *Hopf bifurcation from a differentiable viewpoint*, preprint 1987.

[Kli] A. Klič, *Bifurcations in the systems with involutory symmetry*, preprint 1985.

[Kri&Agl] V.I. Krinsky & K.I. Agladze, *Interaction of rotating waves in an active chemical medium*, Physica **8D** (1983), 50-56.

[Kub&Marek] M. Kubiček & M. Marek, *Computational Methods in Bifurcation Theory and Dissipative Structures*, Springer-Verlag, New York 1983.

[Kü&Mit&Web] T. Küpper & H.D. Mittelmann & H. Weber (eds.), *Numerical Methods for Bifurcation Problems*, Birkhäuser, Basel 1984.

[Kü&Sey&Tro] T. Küpper & R. Seydel & H. Troger (eds.), *Bifurcation: Analysis, Algorithms, Applications*, Inter. Ser. Numer. Math. 79, Birkhäuser, Basel 1987.

[Kun1] P. Kunkel, *Quadratically convergent methods for the computation of unfolded singularities*, to appear in SIAM J. Numer. Anal.

[Kun2] ———— , *Efficient computation of singular points*, preprint 1987.

[Kup] I. Kupka, *Contribution à la théorie des champs génériques*, Contrib. Diff. Eq. **2** (1963), 457-484, and **3** (1964), 411-420.

[Lang] S. Lang, *Algebra*, Addison-Wesley, Reading 1965.

[Las&Y] A. Lasota & J.A. Yorke, *Bounds for periodic solutions of differential equations in Banach spaces*, J. Diff. Eq. **10** (1971), 83-91.

[Lau] R. Lauterbach, *An example of symmetry breaking with submaximal isotropy subgroup*, in [Go&Gu], 217-222.

[M-P&Y1] J. Mallet-Paret & Y.A. Yorke, *Two types of Hopf bifurcation points: sources and sinks of families of periodic orbits*, in: Nonlinear Dynamics, R.H.G. Helleman (ed.), Proc. NY Acad. Sc. **357** (1980), 300-304.

[M-P&Y2] ———— , *Snakes: oriented families of periodic orbits, their sources, sinks, and continuation*, J. Diff. Eq. **43** (1982), 419-450.

[Mars&McCr] J.E. Marsden & M. McCracken, *The Hopf Bifurcation and Its Applications*, Springer-Verlag, New York 1976.

[Mat] T. Matsuoka, *Equivariant function spaces and bifurcation points*, J. Math. Soc. Japan **35** (1983), 43-52.

[Med] M. Medveď, *Generic properties of parametrized vector fields*, I and II, Czech. Math. J. **25** (1975), 376-388, and **26** (1976), 71-83.

[vdM] J.-C. van der Meer, *The Hamiltonian Hopf Bifurcation*, Lect. Notes Math. 1160, Springer-Verlag, Berlin 1985.

[Mit&Web] H.D. Mittelmann & H. Weber, *Bifurcation Problems and their Numerical Solution*, Birkhäuser, Basel 1980.

[Mon&Rob&St] J. Montaldi & M. Roberts & I. Stewart, *Periodic solutions near equilibria of symmetric Hamiltonian systems*, preprint 1986.

[Mü&Ple&Hess] S.C. Müller & T. Plesser & B. Hess, *The structure of the core of the spiral wave in the Belousov-Zhabotinskii reaction*, Science **230** (1985), 661-663.

[Nic&Pri] G. Nicolis & I. Prigogine, *Self-Organization in Nonequilibrium Systems*, Wiley & Sons, New York 1977.

[Oth] H.G. Othmer (ed.), *Nonlinear Oscillations in Biology and Chemistry*, Lect. Notes Biomath. 66, Springer-Verlag, Berlin 1986.

[Oth&Scr] H.G. Othmer & L.E. Scriven, *Instability and dynamic pattern formation in cellular networks*, J. Theor. Biol. **32** (1971), 507-537.

[Pal] J. Palis jr. (ed.), *Geometric Dynamics*, Lect. Notes in Math. 1007, Springer-Verlag, New York 1983.

[Paw&Schm] P.C. Pawlicki & R.A. Schmitz, *Spatial effects on supported catalysts*, Chem. Eng. Progress (1987), 40-45.

[Peit&Prü] H.-O. Peitgen & M. Prüfer, *The Leray-Schauder continuation method is a constructive element in the numerical study of nonlinear eigenvalue and bifurcation problems*, in: Functional Differential Equations and Approximation of Fixed Points, H.-O. Peitgen & H.-O. Walther (eds.), Lect. Notes Math., Springer-Verlag, Berlin 1979.

[Peix] M.M. Peixoto, *On an approximation theorem of Kupka and Smale*, J. Diff. Eq. **3** (1966), 214-227.

[Poi] H. Poincaré, *Les Méthodes Nouvelles de la Mécanique Céleste*, vol. 3, Gauthiers-Villars, Paris 1899.

[Pos1] C. Pospiech, *Global bifurcation with symmetry breaking*, preprint 1986.

[Pos2] ———, *The curious link chain*, in [Kü&Sey&Tro].

[Pos3] ———, *Globale Verzweigung mit Symmetriebrechung für ein Dirichlet-Problem*, Dissertation, Heidelberg 1988.

[Rab] P.H. Rabinowitz, *Some global results for nonlinear eigenvalue problems*, J. Func. Anal. **7** (1971), 487-513.

[Rand] D. Rand, *Dynamics and symmetry: predictions for modulated waves in rotating fluids*, Arch. Rat. Mech. Anal. **79** (1982), 1-38.

[Raz&Schm] L.F. Razón & R.A. Schmitz, *Multiplicities and instabilities in chemically reacting systems – a review*, Chem. Eng. Sci. **42** (1987), 1005-1047.

[Ren] M. Renardy, *Bifurcation from rotating waves*, Arch. Rat. Mech, Anal. **79** (1982), 49-84.

[Ren&Hak] M. Renardy & H. Haken, *Bifurcation of solutions of the Laser equations*, Physica **8D** (1983), 57-89.

[Rhe] W.C. Rheinboldt, *Numerical Analysis of Parametrized Nonlinear Equations*, Wiley & Sons, New York 1986.

[Rob&Swi&Wag] M. Roberts & J.W. Swift & D.H. Wagner, *The Hopf bifurcation on a hexagonal lattice*, in [Go&Gu], 283-318.

[Roo1] D. Roose, *Numerical determination of an emanating branch of Hopf bifurcation points in a two-parameter problem*, preprint 1986.

[Roo2] ———, *Numerical computation of origins for Hopf bifurcation in a two-parameter problem*, in [Kü&Sey&Tro], 268-276.

[Rov] A.B. Rovinsky, *Spiral waves in a model of the ferroin catalyzed Belousov-Zhabotinsky reaction*, J. Chem. Phys. **90** (1986), 217-219.

[Sat1] D.H. Sattinger, *Group Theoretic Methods in Bifurcation Theory*, Lect. Notes Math. 762, Springer-Verlag, Heidelberg 1979.

[Sat2] ———, *Branching in the Presence of Symmetry*, CBMS-NSF Reg. Conf. Ser. Appl. Math. 40, SIAM, Philadelphia 1983.

[Sat&Wea] D.H. Sattinger & O.L. Weaver, *Lie Groups and Algebras with Applications to Physics, Geometry, and Mechanics*, Appl. Math. Sci. 61, Springer-Verlag, New York 1986.

[Sche] S. Schecter, *Bifurcations with symmetry*, in [Mars&McCr], 224-249.

[Schr&Hol&Kub&Marek] I. Schreiber & M. Holodniok & M. Kubiček & M. Marek, *Periodic and aperiodic regimes in coupled dissipative chemical oscillators*, J. Stat. Phys. **43** (1986), 489-519.

[Schr&Marek1] I. Schreiber & M. Marek, *Strange attractors in coupled reaction-diffusion cells*, Physica **5D** (1982), 258-272.

[Schr&Marek2] ———— , *Transition to chaos via two-torus in coupled reaction diffusion cells*, Phys. Lett. **91** (1982), 263-266.

[Ser] J.-P. Serre, *Linear Representations of Finite Groups*, Springer-Verlag, New York 1977.

[Sha] D. Shanks, *Solved and Unsolved Problems in Number Theory*, Chelsea Publ. Comp., New York 1978.

[Šil] L.P. Šilnikov, *A case of the existence of a denumerable set of periodic motions*, Sov. Math. Dokl. **6** (1965), 163-166.

[Sma1] S. Smale, *Stable manifolds for differential equations and diffeomorphisms*, Ann. Sc. Norm. Sup. Pisa **17** (1963), 97-116.

[Sma2] ———— , *An infinite dimensional version of Sard's theorem*, Am. J. Math. **87** (1965), 861-866.

[Sma3] ———— , *A mathematical model of two cells via Turing's equation*, in: Some Mathematical Questions in Biology V, J.D. Cowan (ed.), AMS Lect. Math. Life Sc. **6** (1974), 15-26.

[Smo] J. Smoller, *Shock Waves and Reaction-Diffusion Equations*, Grundl. math. Wiss. 258, Springer-Verlag, New York 1983.

[Smo&Wa1] J. Smoller & A. Wasserman, *Symmetry-breaking for semilinear elliptic equations*, in: Ordinary and Partial Differential Equations, B.D. Sleeman & R.J. Jarvis (eds.), Lect. Notes Math. 1151, Springer-Verlag, New York 1985, 325-334.

[Smo&Wa2] ———— , *Symmetry-breaking for positive solutions of semilinear elliptic equations*, Arch. Rat. Mech. Anal. **95** (1986), 217-225.

[Smo&Wa3] ———— , *Symmetry-breaking for solutions of semilinear elliptic equations with general boundary conditions*, preprint 1986.

[Sot] J. Sotomayor, *Generic bifurcations of dynamical systems*, in: Salvador Symposium on Dynamical Systems, M.M. Peixoto (ed.), Academic Press, New York 1973, 561-582.

[Spa] C. Sparrow, *The Lorenz Equations*, Springer-Verlag, New York 1982.

[Stu&Marek] I. Stuchl & M. Marek, *Dissipative structures in coupled cells: experiments*, J. Chem. Phys. **77** (1982), 2956-2963.

[Swi] J.W. Swift, *Four coupled oscillators: Hopf bifurcation with the symmetry of a square*, preprint 1986.

[Tak] F. Takens, *Singularities of vector fields*, Publ. IHES **43** (1974), 47-100.

[Thom] R. Thom, *Stabilité structurelle et morphogénèse*, Benjamin Inc., Reading 1972.

[Tro] A.J. Tromba, *Fredholm vector fields and a transversality theorem*, J. Func. Anal. **23** (1976), 362-368.

[Tu] A.M. Turing, *The chemical basis of morphogenesis*, Phil. Trans. Roy. Soc. London **B 237** (1952), 37-72.

[Ty1] J.J. Tyson, *The Belousov-Zhabotinskii Reaction*, Lect. Notes Biomath. 10, Springer-Verlag, Berlin 1976.

[Ty2] ———— , *A quantitative account of oscillations, bistability, and traveling waves in the Belousov-Zhabotinskii reaction*, in [Field&Bur], 93-144.

[Van1] A. Vanderbauwhede, *Local Bifurcation and Symmetry*, Pitman, Boston 1982.

[Van2] ———— , *Bifurcation of periodic solutions in a rotationally symmetric oscillation system*, J. reine u. angew. Math. **360** (1985), 1-18.

[Van3] ———— , *Symmetry-breaking at positive solutions*, Z. Angew. Math. Phys. **38** (1987), 315-326.

[Van4] ———— , *Secondary bifurcations of periodic solution in autonomous systems*, CMS annual meeting 1986, W. Langford (ed.), Cont. Math., AMS, Providence, to appear.

[Van5] ———— , *Bifurcation and Symmetry-Breaking*, in preparation.

[vdW] B.L. van der Waerden, *Algebra I*, Springer-Verlag, Berlin 1971.

[Why] G.T. Whyburn, *Topological Analysis*, Princeton Univ. Press, Princeton 1968.

[Win] A.T. Winfree, *Organizing centers for chemical waves in two and three dimensions*, in [Field&Bur], 441-472.

[Y&All] J.A. Yorke & K.T. Alligood, *Cascades of period doublings: a prerequisite for horseshoes*, Bull. AMS **9** (1983), 319-322.

# Subject index

**LECTURE NOTES IN MATHEMATICS**
Edited by A. Dold and B. Eckmann

**Some general remarks on the publication of monographs and seminars**

In what follows all references to monographs, are applicable also to multiauthorship volumes such as seminar notes.

1. Lecture Notes aim to report new developments - quickly, informally, and at a high level. Monograph manuscripts should be reasonably self-contained and rounded off. Thus they may, and often will, present not only results of the author but also related work by other people. Furthermore, the manuscripts should provide sufficient motivation, examples and applications. This clearly distinguishes Lecture Notes manuscripts from journal articles which normally are very concise. Articles intended for a journal but too long to be accepted by most journals, usually do not have this "lecture notes" character. For similar reasons it is unusual for Ph.D. theses to be accepted for the Lecture Notes series.

   Experience has shown that English language manuscripts achieve a much wider distribution.

2. Manuscripts or plans for Lecture Notes volumes should be submitted either to one of the series editors or to Springer-Verlag, Heidelberg. These proposals are then refereed. A final decision concerning publication can only be made on the basis of the complete manuscripts, but a preliminary decision can usually be based on partial information: a fairly detailed outline describing the planned contents of each chapter, and an indication of the estimated length, a bibliography, and one or two sample chapters - or a first draft of the manuscript. The editors will try to make the preliminary decision as definite as they can on the basis of the available information.

3. Lecture Notes are printed by photo-offset from typed copy delivered in camera-ready form by the authors. Springer-Verlag provides technical instructions for the preparation of manuscripts, and will also, on request, supply special staionery on which the prescribed typing area is outlined. Careful preparation of the manuscripts will help keep production time short and ensure satisfactory appearance of the finished book. Running titles are not required; if however they are considered necessary, they should be uniform in appearance. We generally advise authors not to start having their final manuscripts specially tpyed beforehand. For professionally typed manuscripts, prepared on the special stationery according to our instructions, Springer-Verlag will, if necessary, contribute towards the typing costs at a fixed rate.

   The actual production of a Lecture Notes volume takes 6-8 weeks.

.../...

4. Final manuscripts should contain at least 100 pages of mathematical text and should include

   - a table of contents
   - an informative introduction, perhaps with some historical remarks. It should be accessible to a reader not particularly familiar with the topic treated.
   - subject index; this is almost always genuinely helpful for the reader.

5. Authors receive a total of 50 free copies of their volume, but no royalties. They are entitled to purchase further copies of their book for their personal use at a discount of 33 1/3 %, other Springer mathematics books at a discount of 20 % directly from Springer-Verlag.

   Commitment to publish is made by letter of intent rather than by signing a formal contract. Springer-Verlag secures the copyright for each volume.

# LECTURE NOTES

## ESSENTIALS FOR THE PREPARATION OF CAMERA-READY MANUSCRIPTS

The preparation of manuscripts which are to be reproduced by photo-offset requires special care. Manuscripts which are submitted in technically unsuitable form will be returned to the author for retyping. There is normally no possibility of carrying out further corrections after a manuscript is given to production. Hence it is crucial that the following instructions be adhered to closely. If in doubt, please send us 1 - 2 sample pages for examination.

Typing area. On request, Springer-Verlag will supply special paper with the typing area outlined.

The CORRECT TYPING AREA is 18 x 26 1/2 cm (7,5 x 11 inches).

Make sure the TYPING AREA IS COMPLETELY FILLED. Set the margins so that they precisely match the outline and type right from the top to the bottom line. (Note that the page-number will lie outside this area). Lines of text should not end more than three spaces inside or outside the right margin (see example on page 4).

Type on one side of the paper only.

Type. Use an electric typewriter if at all possible. CLEAN THE TYPE before use and always use a BLACK ribbon (a carbon ribbon is best).

Choose a type size large enough to stand reduction to 75%.

Word Processors. Authors using word-processing or computer-typesetting facilities should follow these instructions with obvious modifications. Please note with respect to your printout that
i) the characters should be sharp and sufficiently black;
ii) if the size of your characters is significantly larger or smaller than normal typescript characters, you should adapt the length and breadth of the text area proportionally keeping the proportions 1:0.68.
iii) it is not necessary to use Springer's special typing paper. Any white paper of reasonable quality is acceptable.
IF IN DOUBT, PLEASE SEND US 1-2 SAMPLE PAGES FOR EXAMINATION. We will be glad to give advice.

Spacing and Headings (Monographs). Use ONE-AND-A-HALF line spacing in the text. Please leave sufficient space for the title to stand out clearly and do NOT use a new page for the beginning of subdivisions of chapters. Leave THREE LINES blank above and TWO below headings of such subdivisions.

Spacing and Headings (Proceedings). Use ONE-AND-A-HALF line spacing in the text. Start each paper on a NEW PAGE and leave sufficient space for the title to stand out clearly. However, do NOT use a new page for the beginning of subdivisions of a paper. Leave THREE LINES blank above and TWO below headings of such subdivisions. Make sure headings of equal importance are in the same form.

The first page of each contribution should be prepared in the same way. Therefore, we recommend that the editor prepares a sample page and passes it on to the authors together with these ESSENTIALS. Please take

.../...

the following as an example.

MATHEMATICAL STRUCTURE IN QUANTUM FIELD THEORY

John E. Robert
Fachbereich Physik, Universität Osnabrück
Postfach 44 69, D-4500 Osnabrück

Please leave THREE LINES blank below heading and address of the author. THEN START THE ACTUAL TEXT OF YOUR CONTRIBUTION.

Footnotes. These should be avoided. If they cannot be avoided, place them at the foot of the page, separated from the text by a line 4 cm long, and type them in SINGLE LINE SPACING to finish exactly on the outline.

Symbols. Anything which cannot be typed may be entered by hand in BLACK AND ONLY BLACK ink. (A fine-tipped rapidograph is suitable for this purpose; a good black ball-point will do, but a pencil will not). Do not draw straight lines by hand without a ruler (not even in fractions).

Equations and Computer Programs. Equations and computer programs should begin four spaces inside the left margin. Should the equations be numbered, then each number should be in brackets at the right-hand edge of the typing area.

Pagination. Number pages in the upper right-hand corner in LIGHT BLUE OR GREEN PENCIL ONLY. The final page numbers will be inserted by the printer.

There should normally be NO BLANK PAGES in the manuscript (between chapters or between contributions) unless the book is divided into Part A, Part B for example, which should then begin on a right-hand page.

It is much safer to number pages AFTER the text has been typed and corrected. Page 1 (Arabic) should be THE FIRST PAGE OF THE ACTUAL TEXT. The Roman pagination (table of contents, preface, abstract, acknowledgements, brief introductions, etc.) will be done by Springer-Verlag.

Corrections. When corrections have to be made, cut the new text to fit and PASTE it over the old. White correction fluid may also be used.

Never make corrections or insertions in the text by hand.

If the typescript has to be marked for any reason, e.g. for TEMPORARY page numbers or to mark corrections for the typist, this can be done VERY FAINTLY with BLUE or GREEN PENCIL but NO OTHER COLOR: these colors do not appear after reproduction.

Table of Contents. It is advisable to type the table of contents later, copying the titles from the text and inserting page numbers.

Literature References. These should be placed at the end of each paper or chapter, or at the end of the work, as desired. Type them with single line spacing and start each reference on a new line.
Please ensure that all references are COMPLETE and PRECISE.

Vol. 1145: G. Winkler, Choquet Order and Simplices. VI, 143 pages. 1985.

Vol. 1146: Séminaire d'Algèbre Paul Dubreil et Marie-Paule Malliavin. Proceedings, 1983–1984. Edité par M.-P. Malliavin. IV, 420 pages. 1985.

Vol. 1147: M. Wschebor, Surfaces Aléatoires. VII, 111 pages. 1985.

Vol. 1148: Mark A. Kon, Probability Distributions in Quantum Statistical Mechanics. V, 121 pages. 1985.

Vol. 1149: Universal Algebra and Lattice Theory. Proceedings, 1984. Edited by S. D. Comer. VI, 282 pages. 1985.

Vol. 1150: B. Kawohl, Rearrangements and Convexity of Level Sets in PDE. V, 136 pages. 1985.

Vol 1151: Ordinary and Partial Differential Equations. Proceedings, 1984. Edited by B.D. Sleeman and R.J. Jarvis. XIV, 357 pages. 1985.

Vol. 1152: H. Widom, Asymptotic Expansions for Pseudodifferential Operators on Bounded Domains. V, 150 pages. 1985.

Vol. 1153: Probability in Banach Spaces V. Proceedings, 1984. Edited by A. Beck, R. Dudley, M. Hahn, J. Kuelbs and M. Marcus. VI, 457 pages. 1985.

Vol. 1154: D.S. Naidu, A.K. Rao, Singular Pertubation Analysis of Discrete Control Systems. IX, 195 pages. 1985.

Vol. 1155: Stability Problems for Stochastic Models. Proceedings, 1984. Edited by V.V. Kalashnikov and V.M. Zolotarev. VI, 447 pages. 1985.

Vol. 1156: Global Differential Geometry and Global Analysis 1984. Proceedings, 1984. Edited by D. Ferus, R.B. Gardner, S. Helgason and U. Simon. V, 339 pages. 1985.

Vol. 1157: H. Levine, Classifying Immersions into $\mathbb{R}^4$ over Stable Maps of 3-Manifolds into $\mathbb{R}^2$. V, 163 pages. 1985.

Vol. 1158: Stochastic Processes – Mathematics and Physics. Proceedings, 1984. Edited by S. Albeverio, Ph. Blanchard and L. Streit. VI, 230 pages. 1986.

Vol. 1159: Schrödinger Operators, Como 1984. Seminar. Edited by S. Graffi. VIII, 272 pages. 1986.

Vol. 1160: J.-C. van der Meer, The Hamiltonian Hopf Bifurcation. VI, 115 pages. 1985.

Vol. 1161: Harmonic Mappings and Minimal Immersions, Montecatini 1984. Seminar. Edited by E. Giusti. VII, 285 pages. 1985.

Vol. 1162: S.J.L. van Eijndhoven, J. de Graaf, Trajectory Spaces, Generalized Functions and Unbounded Operators. IV, 272 pages. 1985.

Vol. 1163: Iteration Theory and its Functional Equations. Proceedings, 1984. Edited by R. Liedl, L. Reich and Gy. Targonski. VIII, 231 pages. 1985.

Vol. 1164: M. Meschiari, J.H. Rawnsley, S. Salamon, Geometry Seminar "Luigi Bianchi" II – 1984. Edited by E. Vesentini. VI, 224 pages. 1985.

Vol. 1165: Seminar on Deformations. Proceedings, 1982/84. Edited by J. Ławrynowicz. IX, 331 pages. 1985.

Vol. 1166: Banach Spaces. Proceedings, 1984. Edited by N. Kalton and E. Saab. VI, 199 pages. 1985.

Vol. 1167: Geometry and Topology. Proceedings, 1983–84. Edited by J. Alexander and J. Harer. VI, 292 pages. 1985.

Vol. 1168: S.S. Agaian, Hadamard Matrices and their Applications. III, 227 pages. 1985.

Vol. 1169: W.A. Light, E.W. Cheney, Approximation Theory in Tensor Product Spaces. VII, 157 pages. 1985.

Vol. 1170: B.S. Thomson, Real Functions. VII, 229 pages. 1985.

Vol. 1171: Polynômes Orthogonaux et Applications. Proceedings, 1984. Edité par C. Brezinski, A. Draux, A.P. Magnus, P. Maroni et A. Ronveaux. XXXVII, 584 pages. 1985.

Vol. 1172: Algebraic Topology, Göttingen 1984. Proceedings. Edited by L. Smith. VI, 209 pages. 1985.

Vol. 1173: H. Delfs, M. Knebusch, Locally Semialgebraic Spaces. XVI, 329 pages. 1985.

Vol. 1174: Categories in Continuum Physics, Buffalo 1982. Seminar. Edited by F.W. Lawvere and S.H. Schanuel. V, 126 pages. 1986.

Vol. 1175: K. Mathiak, Valuations of Skew Fields and Projective Hjelmslev Spaces. VII, 116 pages. 1986.

Vol. 1176: R.R. Bruner, J.P. May, J.E. McClure, M. Steinberger, $H_\infty$ Ring Spectra and their Applications. VII, 388 pages. 1986.

Vol. 1177: Representation Theory I. Finite Dimensional Algebras. Proceedings, 1984. Edited by V. Dlab, P. Gabriel and G. Michler. XV, 340 pages. 1986.

Vol. 1178: Representation Theory II. Groups and Orders. Proceedings, 1984. Edited by V. Dlab, P. Gabriel and G. Michler. XV, 370 pages. 1986.

Vol. 1179: Shi J.-Y. The Kazhdan-Lusztig Cells in Certain Affine Weyl Groups. X, 307 pages. 1986.

Vol. 1180: R. Carmona, H. Kesten, J.B. Walsh, École d'Été de Probabilités de Saint-Flour XIV – 1984. Édité par P.L. Hennequin. X, 438 pages. 1986.

Vol. 1181: Buildings and the Geometry of Diagrams, Como 1984. Seminar. Edited by L. Rosati. VII, 277 pages. 1986.

Vol. 1182: S. Shelah, Around Classification Theory of Models. VII, 279 pages. 1986.

Vol. 1183: Algebra, Algebraic Topology and their Interactions. Proceedings, 1983. Edited by J.-E. Roos. XI, 396 pages. 1986.

Vol. 1184: W. Arendt, A. Grabosch, G. Greiner, U. Groh, H.P. Lotz, U. Moustakas, R. Nagel, F. Neubrander, U. Schlotterbeck, One-parameter Semigroups of Positive Operators. Edited by R. Nagel. X, 460 pages. 1986.

Vol. 1185: Group Theory, Beijing 1984. Proceedings. Edited by Tuan H.F. V, 403 pages. 1986.

Vol. 1186: Lyapunov Exponents. Proceedings, 1984. Edited by L. Arnold and V. Wihstutz. VI, 374 pages. 1986.

Vol. 1187: Y. Diers, Categories of Boolean Sheaves of Simple Algebras. VI, 168 pages. 1986.

Vol. 1188: Fonctions de Plusieurs Variables Complexes V. Séminaire, 1979–85. Edité par François Norguet. VI, 306 pages. 1986.

Vol. 1189: J. Lukeš, J. Malý, L. Zajíček, Fine Topology Methods in Real Analysis and Potential Theory. X, 472 pages. 1986.

Vol. 1190: Optimization and Related Fields. Proceedings, 1984. Edited by R. Conti, E. De Giorgi and F. Giannessi. VIII, 419 pages. 1986.

Vol. 1191: A.R. Its, V.Yu. Novokshenov, The Isomonodromic Deformation Method in the Theory of Painlevé Equations. IV, 313 pages. 1986.

Vol. 1192: Equadiff 6. Proceedings, 1985. Edited by J. Vosmansky and M. Zlámal. XXIII, 404 pages. 1986.

Vol. 1193: Geometrical and Statistical Aspects of Probability in Banach Spaces. Proceedings, 1985. Edited by X. Fernique, B. Heinkel, M.B. Marcus and P.A. Meyer. IV, 128 pages. 1986.

Vol. 1194: Complex Analysis and Algebraic Geometry. Proceedings, 1985. Edited by H. Grauert. VI, 235 pages. 1986.

Vol.1195: J.M. Barbosa, A.G. Colares, Minimal Surfaces in $\mathbb{R}^3$. X, 124 pages. 1986.

Vol. 1196: E. Casas-Alvero, S. Xambó-Descamps, The Enumerative Theory of Conics after Halphen. IX, 130 pages. 1986.

Vol. 1197: Ring Theory. Proceedings, 1985. Edited by F.M.J. van Oystaeyen. V, 231 pages. 1986.

Vol. 1198: Séminaire d'Analyse, P. Lelong – P. Dolbeault – H. Skoda. Seminar 1983/84. X, 260 pages. 1986.

Vol. 1199: Analytic Theory of Continued Fractions II. Proceedings, 1985. Edited by W.J. Thron. VI, 299 pages. 1986.

Vol. 1200: V.D. Milman, G. Schechtman, Asymptotic Theory of Finite Dimensional Normed Spaces. With an Appendix by M. Gromov. VIII, 156 pages. 1986.

Vol. 1201: Curvature and Topology of Riemannian Manifolds. Proceedings, 1985. Edited by K. Shiohama, T. Sakai and T. Sunada. VII, 336 pages. 1986.

Vol. 1202: A. Dür, Möbius Functions, Incidence Algebras and Power Series Representations. XI, 134 pages. 1986.

Vol. 1203: Stochastic Processes and Their Applications. Proceedings, 1985. Edited by K. Itô and T. Hida. VI, 222 pages. 1986.

Vol. 1204: Séminaire de Probabilités XX, 1984/85. Proceedings. Edité par J. Azéma et M. Yor. V, 639 pages. 1986.

Vol. 1205: B.Z. Moroz, Analytic Arithmetic in Algebraic Number Fields. VII, 177 pages. 1986.

Vol. 1206: Probability and Analysis, Varenna (Como) 1985. Seminar. Edited by G. Letta and M. Pratelli. VIII, 280 pages. 1986.

Vol. 1207: P.H. Bérard, Spectral Geometry: Direct and Inverse Problems. With an Appendix by G. Besson. XIII, 272 pages. 1986.

Vol. 1208: S. Kaijser, J.W. Pelletier, Interpolation Functors and Duality. IV, 167 pages. 1986.

Vol. 1209: Differential Geometry, Peñíscola 1985. Proceedings. Edited by A.M. Naveira, A. Ferrández and F. Mascaró. VIII, 306 pages. 1986.

Vol. 1210: Probability Measures on Groups VIII. Proceedings, 1985. Edited by H. Heyer. X, 386 pages. 1986.

Vol. 1211: M.B. Sevryuk, Reversible Systems. V, 319 pages. 1986.

Vol. 1212: Stochastic Spatial Processes. Proceedings, 1984. Edited by P. Tautu. VIII, 311 pages. 1986.

Vol. 1213: L.G. Lewis, Jr., J.P. May, M. Steinberger, Equivariant Stable Homotopy Theory. IX, 538 pages. 1986.

Vol. 1214: Global Analysis – Studies and Applications II. Edited by Yu.G. Borisovich and Yu.E. Gliklikh. V, 275 pages. 1986.

Vol. 1215: Lectures in Probability and Statistics. Edited by G. del Pino and R. Rebolledo. V, 491 pages. 1986.

Vol. 1216: J. Kogan, Bifurcation of Extremals in Optimal Control. VIII, 106 pages. 1986.

Vol. 1217: Transformation Groups. Proceedings, 1985. Edited by S. Jackowski and K. Pawalowski. X, 396 pages. 1986.

Vol. 1218: Schrödinger Operators, Aarhus 1985. Seminar. Edited by E. Balslev. V, 222 pages. 1986.

Vol. 1219: R. Weissauer, Stabile Modulformen und Eisensteinreihen. III, 147 Seiten. 1986.

Vol. 1220: Séminaire d'Algèbre Paul Dubreil et Marie-Paule Malliavin. Proceedings, 1985. Edité par M.-P. Malliavin. IV, 200 pages. 1986.

Vol. 1221: Probability and Banach Spaces. Proceedings, 1985. Edited by J. Bastero and M. San Miguel. XI, 222 pages. 1986.

Vol. 1222: A. Katok, J.-M. Strelcyn, with the collaboration of F. Ledrappier and F. Przytycki, Invariant Manifolds, Entropy and Billiards; Smooth Maps with Singularities. VIII, 283 pages. 1986.

Vol. 1223: Differential Equations in Banach Spaces. Proceedings, 1985. Edited by A. Favini and E. Obrecht. VIII, 299 pages. 1986.

Vol. 1224: Nonlinear Diffusion Problems, Montecatini Terme 1985. Seminar. Edited by A. Fasano and M. Primicerio. VIII, 188 pages. 1986.

Vol. 1225: Inverse Problems, Montecatini Terme 1986. Seminar. Edited by G. Talenti. VIII, 204 pages. 1986.

Vol. 1226: A. Buium, Differential Function Fields and Moduli of Algebraic Varieties. IX, 146 pages. 1986.

Vol. 1227: H. Helson, The Spectral Theorem. VI, 104 pages. 1986.

Vol. 1228: Multigrid Methods II. Proceedings, 1985. Edited by W. Hackbusch and U. Trottenberg. VI, 336 pages. 1986.

Vol. 1229: O. Bratteli, Derivations, Dissipations and Group Actions on C*-algebras. IV, 277 pages. 1986.

Vol. 1230: Numerical Analysis. Proceedings, 1984. Edited by J.-P. Hennart. X, 234 pages. 1986.

Vol. 1231: E.-U. Gekeler, Drinfeld Modular Curves. XIV, 107 pages. 1986.

Vol. 1232: P.C. Schuur, Asymptotic Analysis of Soliton Problems. VIII, 180 pages. 1986.

Vol. 1233: Stability Problems for Stochastic Models. Proceedings, 1985. Edited by V.V. Kalashnikov, B. Penkov and V.M. Zolotarev. VI, 223 pages. 1986.

Vol. 1234: Combinatoire énumérative. Proceedings, 1985. Edité par G. Labelle et P. Leroux. XIV, 387 pages. 1986.

Vol. 1235: Séminaire de Théorie du Potentiel, Paris, No. 8. Directeurs: M. Brelot, G. Choquet et J. Deny. Rédacteurs: F. Hirsch et G. Mokobodzki. III, 209 pages. 1987.

Vol. 1236: Stochastic Partial Differential Equations and Applications. Proceedings, 1985. Edited by G. Da Prato and L. Tubaro. V, 257 pages. 1987.

Vol. 1237: Rational Approximation and its Applications in Mathematics and Physics. Proceedings, 1985. Edited by J. Gilewicz, M. Pindor and W. Siemaszko. XII, 350 pages. 1987.

Vol. 1238: M. Holz, K.-P. Podewski and K. Steffens, Injective Choice Functions. VI, 183 pages. 1987.

Vol. 1239: P. Vojta, Diophantine Approximations and Value Distribution Theory. X, 132 pages. 1987.

Vol. 1240: Number Theory, New York 1984–85. Seminar. Edited by D.V. Chudnovsky, G.V. Chudnovsky, H. Cohn and M.B. Nathanson. V, 324 pages. 1987.

Vol. 1241: L. Gårding, Singularities in Linear Wave Propagation. III, 125 pages. 1987.

Vol. 1242: Functional Analysis II, with Contributions by J. Hoffmann-Jørgensen et al. Edited by S. Kurepa, H. Kraljević and D. Butković. VII, 432 pages. 1987.

Vol. 1243: Non Commutative Harmonic Analysis and Lie Groups. Proceedings, 1985. Edited by J. Carmona, P. Delorme and M. Vergne. V, 309 pages. 1987.

Vol. 1244: W. Müller, Manifolds with Cusps of Rank One. XI, 158 pages. 1987.

Vol. 1245: S. Rallis, L-Functions and the Oscillator Representation. XVI, 239 pages. 1987.

Vol. 1246: Hodge Theory. Proceedings, 1985. Edited by E. Cattani, F. Guillén, A. Kaplan and F. Puerta. VII, 175 pages. 1987.

Vol. 1247: Séminaire de Probabilités XXI. Proceedings. Edité par J. Azéma, P.A. Meyer et M. Yor. IV, 579 pages. 1987.

Vol. 1248: Nonlinear Semigroups, Partial Differential Equations and Attractors. Proceedings, 1985. Edited by T.L. Gill and W.W. Zachary. IX, 185 pages. 1987.

Vol. 1249: I. van den Berg, Nonstandard Asymptotic Analysis. IX, 187 pages. 1987.

Vol. 1250: Stochastic Processes – Mathematics and Physics II. Proceedings 1985. Edited by S. Albeverio, Ph. Blanchard and L. Streit. VI, 359 pages. 1987.

Vol. 1251: Differential Geometric Methods in Mathematical Physics. Proceedings, 1985. Edited by P.L. García and A. Pérez-Rendón. VII, 300 pages. 1987.

Vol. 1252: T. Kaise, Représentations de Weil et $GL_2$ Algèbres de division et $GL_n$. VII, 203 pages. 1987.

Vol. 1253: J. Fischer, An Approach to the Selberg Trace Formula via the Selberg Zeta-Function. III, 184 pages. 1987.

Vol. 1254: S. Gelbart, I. Piatetski-Shapiro, S. Rallis. Explicit Constructions of Automorphic L-Functions. VI, 152 pages. 1987.

Vol. 1255: Differential Geometry and Differential Equations. Proceedings, 1985. Edited by C. Gu, M. Berger and R.L. Bryant. XII, 243 pages. 1987.

Vol. 1256: Pseudo-Differential Operators. Proceedings, 1986. Edited by H.O. Cordes, B. Gramsch and H. Widom. X, 479 pages. 1987.

Vol. 1257: X. Wang, On the C*-Algebras of Foliations in the Plane. V, 165 pages. 1987.

Vol. 1258: J. Weidmann, Spectral Theory of Ordinary Differential Operators. VI, 303 pages. 1987.